RESOURCING DEFENSE COOPERATION IN EUROPE
AMIDST RUSSIA'S WAR IN UKRAINE

Implications of Reforms to Planning, Programming, Budgeting, and Execution Processes for Multinational Collaboration

JAMES BLACK | STUART DEE | MEGAN McKERNAN

STEPHANIE YOUNG | NICOLAS JOUAN

MAXIME SOMMERFELD ANTONIOU

CLARA LE GARGASSON | LINDA SLAPAKOVA

THEODORA OGDEN | ERIK SILFVERSTEN | MATTIAS EKEN

CHARLOTTE KLEBERG | JOHN P. GODGES

Prepared for the Commission on Planning, Programming, Budgeting, and Execution Reform

NATIONAL DEFENSE RESEARCH INSTITUTE

For more information on this publication, visit **www.rand.org/t/RRA2195-8**.

About RAND

The RAND Corporation is a research organization that develops solutions to public policy challenges to help make communities throughout the world safer and more secure, healthier and more prosperous. RAND is nonprofit, nonpartisan, and committed to the public interest. To learn more about RAND, visit www.rand.org.

Research Integrity

Our mission to help improve policy and decisionmaking through research and analysis is enabled through our core values of quality and objectivity and our unwavering commitment to the highest level of integrity and ethical behavior. To help ensure our research and analysis are rigorous, objective, and nonpartisan, we subject our research publications to a robust and exacting quality-assurance process; avoid both the appearance and reality of financial and other conflicts of interest through staff training, project screening, and a policy of mandatory disclosure; and pursue transparency in our research engagements through our commitment to the open publication of our research findings and recommendations, disclosure of the source of funding of published research, and policies to ensure intellectual independence. For more information, visit www.rand.org/about/principles.

RAND's publications do not necessarily reflect the opinions of its research clients and sponsors.

Published by the RAND Corporation, Santa Monica, Calif.
© 2024 RAND Corporation
RAND® is a registered trademark.

Library of Congress Cataloging-in-Publication Data is available for this publication.

ISBN: 978-1-9774-1353-6

Cover image by Airman 1st Class Cydney Lee/U.S. Air Force and vika_k/Adobe Stock.

Limited Print and Electronic Distribution Rights

About This Report

The analysis presented in this report addresses the implications of changing U.S. and European approaches to planning, programming, budgeting, and execution (PPBE) processes in national defense organizations, along with the associated challenges and opportunities for collaboration on joint capability development and acquisition programs between North Atlantic Treaty Organization (NATO) Allies. Although the term *PPBE* is derived from the U.S. Department of Defense's PPBE System, we use it generically in this report to describe how various entities (i.e., national governments, the European Union [EU], and NATO) align their strategies and plans to resource their joint defense.

This report should be of interest to both U.S. and European policymakers and scholars who want a snapshot of current European and NATO defense collaboration efforts through the lens of PPBE. The research reported here was completed in March 2024 and underwent security review with the sponsor and the Defense Office of Prepublication and Security Review before public release.

RAND National Security Research Division

This research was sponsored by the Commission on PPBE Reform and conducted within the Acquisition and Technology Policy (ATP) Program of the RAND National Security Research Division (NSRD), which operates the National Defense Research Institute (NDRI), a federally funded research and development center sponsored by the Office of the Secretary of Defense, the Joint Staff, the Unified Combatant Commands, the Navy, the Marine Corps, the defense agencies, and the defense intelligence enterprise.

For more information on the RAND ATP Program, see www.rand.org/nsrd/atp or contact the director (contact information is provided on the webpage).

RAND Europe

This research was supported by RAND Europe. RAND Europe has offices in the United Kingdom, Belgium, and the Netherlands and conducts research and analysis for European governments, as well as EU and NATO institutions.

Acknowledgments

We thank the members of the Commission on PPBE Reform—Robert Hale, Ellen Lord, Jonathan Burks, Susan Davis, Lisa Disbrow, Eric Fanning, Peter Levine, Jamie Morin, David Norquist, Diem Salmon, Jennifer Santos, Arun Seraphin, Raj Shah, and John Whitley—and staff for their dedication and deep expertise in shaping this work. We extend special gratitude to the commission chair, the Honorable Robert Hale; the vice chair, the Honorable Ellen Lord; executive director Lara Sayer; director of research Elizabeth Bieri; and the commission's senior researcher Kelle McCluskey for their guidance and support throughout this analysis. We would also like to thank the subject-matter experts on France, Germany, Sweden, and the United Kingdom who provided us valuable insight.

From NSRD, we thank Barry Pavel, vice president and director, and Caitlin Lee, ATP director. We also thank our peer reviewers—Hans Pung and Christian Trotti—who offered helpful feedback on this report. Finally, we would like to thank Maria Falvo and Saci Haslam for their administrative assistance. The work is much improved by their inputs, but any errors remain the responsibility of the authors alone.

Summary

This report explores the deepening of European defense cooperation against the backdrop of Russia's war of aggression in Ukraine. Since the full-scale invasion in February 2022, Europe, the United States, and other democratic nations have mobilized large amounts of financial and military aid to support Kyiv. Crucially, as the war has gone on, Ukraine's supporters have recognized the importance of not just providing more aid but also improving the coordination of providing said aid. More generally, defense organizations on both sides of the Atlantic have benefited from increases in national defense spending and have sought to find novel ways of funding and procuring new equipment and munitions at pace. Pursuing efficiency and economies of scale, the European Union (EU) and the North Atlantic Treaty Organization (NATO) have both introduced plans and measures to increase military and industrial capability, capacity, and readiness. These are prerequisites for a more robust collective defense and deterrence posture in Europe.

Part of a wider body of RAND research into prospective reforms of the U.S. Department of Defense's planning, programming, budgeting, and execution (PPBE) processes,[1] this report examines the role that allied PPBE-like mechanisms play in either enabling or frustrating collaboration across the EU, NATO, and the United States amid a deteriorating threat environment:

- First, **we explore selected case studies of evolving national approaches to defense resourcing** in the United Kingdom, France, Germany, and Sweden.
- Second, **we examine the implications of the major initiatives recently launched by the EU** (e.g., the European Defence Fund, the Act in Support of Ammunition Production, the European Defence Industry Reinforcement Through Common Procurement Act, and the first-ever European Defence Industrial Strategy [EDIS]) **and by NATO** (e.g., the Defence Production Action Plan [DPAP] and a new Innovation Fund, called the Defence Innovation Accelerator for the North Atlantic), as well as U.S. financial instruments open to European allies.

---

[1] Megan McKernan, Stephanie Young, Andrew Dowse, James Black, Devon Hill, Benjamin J. Sacks, Austin Wyatt, Nicolas Jouan, Yuliya Shokh, Jade Yeung, Raphael S. Cohen, John P. Godges, Heidi Peters, and Lauren Skrabala, *Planning, Programming, Budgeting, and Execution in Comparative Organizations: Vol. 2, Case Studies of Selected Allied and Partner Nations*, RAND Corporation, RR-A2195-2, 2024; Stephanie Young, Megan McKernan, Andrew Dowse, Nicolas Jouan, Theodora Ogden, Austin Wyatt, Mattias Eken, Linda Slapakova, Naoko Aoki, Clara Le Gargasson, Charlotte Kleberg, Maxime Sommerfeld Antoniou, Phoebe Felicia Pham, Jade Yeung, Turner Ruggi, Erik Silfversten, James Black, Raphael S. Cohen, John P. Godges, Heidi Peters, and Lauren Skrabala, *Planning, Programming, Budgeting, and Execution in Comparative Organizations: Vol. 5, Additional Case Studies of Selected Allied and Partner Nations*, RAND Corporation, RR-A2195-5, 2024.

- Last, **we assess progress made toward enabling and incentivizing cooperation** through reforms to national or collective PPBE processes.

Our analysis shows that the war in Ukraine has provided fresh impetus to continue prewar trends toward the reform of and pursuit of greater coherence between various actors' PPBE processes as basic enablers of defense cooperation.

Important progress has been made on several fronts. Recent initiatives, such as the capability coalitions assembled to support Ukraine, the EU's EDIS and associated new financial mechanisms, and NATO's DPAP, all help to enable and incentivize a more efficient collective use of finite financial, military, and industrial resources across the transatlantic community. At the same time, significant barriers and countervailing pressures persist, not least of which is the enduring focus of most national governments on sovereign programs, support to domestic industries, and the prioritization of short- over long-term goals.

Further action—backed by the political will to make compromises in pursuit of collective benefits—is needed if the EU and NATO are to more efficiently resource their strategic ambitions: more-robust defense and deterrence, more-resilient industrial base and supply chains, a more conducive environment for innovation, and, ultimately, a West better positioned to deal with both war in Ukraine and other emerging threats, such as China.

Contents

Figures and Tables

Figures

Tables

CHAPTER 1

Drivers of, and Barriers to, European Defense Cooperation

We begin by discussing the drivers of, and barriers to, deepening European defense cooperation against the backdrop of Russia's war of aggression in Ukraine. We then consider the specific role of planning, programming, budgeting, and execution (PPBE) processes in enabling or frustrating such collaboration, building on extensive, recent RAND analyses in support of the Commission on PPBE Reform.

Russia's full-scale invasion of Ukraine in February 2022 brought a renewed focus on threats to European territorial defense and the wider collective security of North Atlantic Treaty Organization (NATO) Allies. The United States, Canada, and their European allies have donated significant financial and military aid to Ukraine, as well as taken in millions of Ukrainian refugees. NATO has also accepted Finland and Sweden into the Alliance and adopted a new force model and deterrence posture to mitigate the increased Russian threat, surging forces to NATO's eastern flank.

All such efforts have stretched defense budgets and strained industrial capacity on both sides of the Atlantic, but especially in Europe, where the defense industry experienced decades of low investment following the post–Cold War "peace dividend." Countries have worked together to coordinate the provision of aid to Kyiv, using European Union (EU), NATO, and other ad hoc mechanisms to jointly procure artillery rounds, missiles, and other materiel. Most European nations have increased their defense spending—in some cases, very sharply— to backfill stockpiles of equipment and munitions donated to Ukraine, as well as to expand their own capacity to deter and defend against direct threats to the rest of Europe.

Still, European and, by extension, NATO efforts remain fragmented. European nations and other NATO Allies vary significantly in terms of their economies, armed forces, financial resources, and underlying political systems. They collectively represent a wide array of defense industrial capabilities and bring to bear significant spending potential, but they continue to be held back by a fragmented market and diffuse approaches to procurement. It is not only that every European nation, except Poland,[1] spends a smaller percentage of its gross domestic product (GDP) on defense than the United States does (although defense spending

---

[1] As of 2023, Poland was estimated to spend nearly 4 percent of its GDP on defense (NATO, "The Secretary General's Annual Report 2023," webpage, March 14, 2024f).

in Europe has increased sharply since February 2022 and will reach an average of 2 percent in 2024).[2] European nations also tend to get less value from what they do invest in defense: A fragmented approach to spending makes it harder to generate beneficial economies of scale in terms of research and development (R&D), production, procurement, and capability life-cycle support costs. Varying degrees of political buy-in to deeper EU or NATO integration complicate prospects for the more efficient use of resources still further, as do enduring barriers to alignment with the United States, including via its Foreign Military Sales (FMS) program.

As a result, European armed forces, taken together, operate a much wider variety of equipment types within each category of materiel (e.g., main battle tanks, artillery systems, frigates) than the U.S. military does. This fragmentation has negative consequences for life-cycle costs, industrial competitiveness, workforce skills, investment, exports, and interoperability. While some European-manufactured systems are cutting edge, many fail to compete with the best that the United States has to offer. This quality difference is exacerbated by the fragmented nature of both the European defence technological and industrial base (EDTIB) and the European defence equipment market (EDEM). Such inefficiencies, in turn, frustrate efforts to promote innovation and to develop and roll out disruptive new products and services at scale.

Over the past few decades, various initiatives have sought to address these issues by deepening cooperation on defense resourcing, capability development, acquisition, and industrial policy. For some countries, especially in Europe, this cooperation represents a political end in and of itself—in line with the federalist or integrationist ambitions of the EU more generally and the spirit of the single market. Many others, however, see cooperation more as a means to an end, not least because such collaborative programs can introduce new costs, compromises, and complexities compared with acting alone and, thus, must be justified in terms of their benefits (whether politically, militarily, or economically) compared with sovereign action.[3]

Regardless of the underlying policy motivations, defense cooperation has emerged as a prominent way in which like-minded allies try to more efficiently use finite resources by pooling and sharing them—a strategy of particular relevance to European nations, which lack anything like the economic or military clout of the United States when working individually but collectively have access to a larger GDP and population than their U.S. ally. Recent years have seen successive waves of EU-led efforts to bolster European strategic autonomy, promote more multinational procurements, increase intra-EU defense trade, and encourage consolidation of national defense industries.

---

[2] NATO, "Secretary General Welcomes Unprecedented Rise in NATO Defence Spending," February 15, 2024d.

[3] Indeed, as we discuss later, there remains a significant say-do gap between national governments' repeated public statements in favor of cooperation and the realities of their activities and investments, which continue to focus primarily on sovereign action and on national priorities, such as maintaining local jobs, industrial security of supply, or access to certain technologies. Thus, there is a constant tension between elements of (friendly) competition and collaboration in national approaches to defense, even among the closest allies.

Although the EU is growing in confidence as a defense and security actor, NATO remains Europe's primary collective security alliance. Within this context, the United States continues to significantly outspend its European allies on defense capability, equipment, and R&D, at around 70 percent of the total combined NATO expenditure. Furthermore, U.S. spending not only outstrips that of Allies in absolute terms (which is unsurprising given that the United States has comfortably the largest population or economy of any single NATO member country). The United States also leads in terms of meeting NATO's targets for its members to spend more than 2 percent of GDP on defense and more than 20 percent of their defense budgets on new equipment (as opposed to spending on such costs as personnel), as shown in Figure 1.1.

The landscape of defense cooperation mechanisms in Europe is highly complex, with collaborative mechanisms varying in intent, ambition, and membership, as shown in Figure 1.2. Besides the EU and NATO, this landscape includes initiatives for developing new military and industrial capabilities, such as Permanent Structured Cooperation (PESCO); bodies for delivering joint acquisition programs, such as the Organisation for Joint Armament Cooperation (OCCAR); more–operationally focused groupings, such as the United Kingdom (UK)-led Joint Expeditionary Force (JEF) or the European Intervention Initiative (EI2); and regionally focused or minilateral clusters, such as the Nordic Defence Cooperation (NORDEFCO), the Northern Group, the Visegrad Group, or the UK's participation in the Australia–United Kingdom–United States (AUKUS) partnership.[4]

These initiatives also involve varying levels of integration with non-EU countries, such as the UK, Norway, or Ukraine, or with non-European nations, most notably the United States. Many nations also have bilateral treaties in place, such as the Defense Cooperation Agreements signed between the United States and each of the following countries: Bulgaria, Czechia, Estonia, Hungary, Latvia, Lithuania, Norway, Poland, Slovakia, and Sweden (at the time of this writing, negotiations with Denmark had recently concluded). It is this complex and inefficient web of collaborative arrangements that the United States and its European allies must navigate when seeking to deliver on their stated policy ambitions for deeper defense cooperation, including the development and acquisition of new military capabilities to defend NATO or the ramping up of financial and military support to Ukraine.

In practice, collaboration on fielding new defense capabilities has a checkered history and has yielded sporadic success. As the European Commission notes, defense acquisitions in the EU are characterized by "buying predominantly alone and from abroad."[5] This assessment would equally apply well to most non-EU nations in Europe, such as Norway

---

[4] For more on the impact of U.S., UK, and Australian PPBE processes on AUKUS, see Andrew Dowse, Megan McKernan, James Black, Stephanie Young, Austin Wyatt, John P. Godges, Nicolas Jouan, and Joanne Nicholson, *AUKUS Collaboration Throughout the Capability Life Cycle: Implications for Planning, Programming, Budgeting, and Execution Processes*, RAND Corporation, PE-A2195-1, 2024.

[5] European Commission, "A New European Defence Industrial Strategy: Achieving EU Readiness Through a Responsive and Resilient European Defence Industry," joint communication to the European Parliament, the Council, the European Economic and Social Committee, and the Committee of the Regions, March 3, 2024a.

FIGURE 1.1

NATO Allies' Financial Contributions in Context

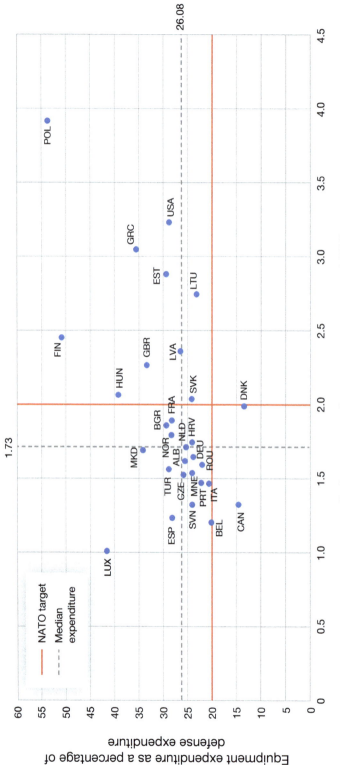

SOURCE: Adapted from NATO, 2024f.

NOTE: Data relate to estimates for national defense expenditure in 2023. ALB = Albania; BEL = Belgium; BGR = Bulgaria; CAN = Canada; CZE = Czechia; DEU = Germany; DNK = Denmark; ESP = Spain; EST = Estonia; FIN = Finland; FRA = France; GBR = UK; GRC = Greece; HRV = Croatia; HUN = Hungary; ITA = Italy; LTU = Lithuania; LUX = Luxembourg; LVA = Latvia; MKD = North Macedonia; MNE = Montenegro; NLD = Netherlands; NOR = Norway; POL = Poland; PRT = Portugal; ROU = Romania; SVK = Slovakia; SVN = Slovenia; TUR = Türkiye; USA = United States.

FIGURE 1.2
Selection of Overlapping European Defense Cooperation Initiatives

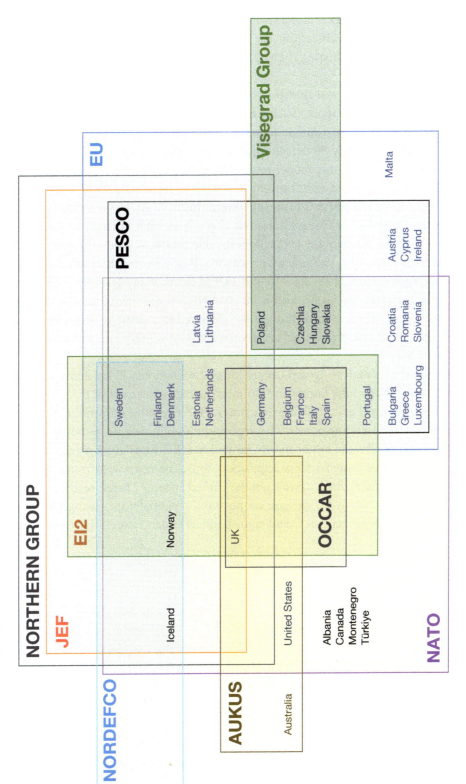

or the UK. Most countries retain a strong focus on sovereignty, insist on bespoke requirements and technical specifications for new equipment, and seek to procure equipment either from their own "national champions" in industry—if they exist—or from non-European suppliers.[6]

The emphasis on national champions reflects both economic protectionism and concern for military security of supply. (Such trends are also present in the industrial, acquisition, and export policies of the United States.) The European Commission has previously estimated that approximately 80 percent of defense contracts across the EU are being placed with national suppliers; this model results in significant duplication. For example, there were 19 separate armored fighting vehicle programs reportedly underway across the EU prior to the full-scale Russian invasion of Ukraine in 2022.[7] At the same time, non-European suppliers (principally the United States but also others, such as Israel and the Republic of Korea) play a vital role in many of the highest-value contracts, especially for fighter aircraft, helicopters, air defenses, and precision weapons. Foreign companies are aided by the fact that the fragmentation of the EDTIB and EDEM leaves many European firms struggling to achieve the resources, scale, and associated efficiencies needed to stay truly competitive. Procurements from outside Europe also often reflect a desire to access new technology, encourage inward investment (including via offsets), or solidify political relationships with the United States.

Overall, then, progress toward deeper integration between EU member states has been limited.[8] According to the European Commission, only 18 percent of EU members' defense procurement budgets go to collaborative programs.[9] The value of intra-EU defense trade is similarly low and trending poorly. The urgent ramp-up in defense spending since the full-scale invasion of Ukraine has primarily favored non-EU suppliers: Between February 2022 and June 2023, 78 percent of defense acquisitions by EU members were from countries outside the EU—63 percent of which came from the United States.[10] Therefore, the war in Ukraine has "negatively impacted EU defense cooperation, potentially increasing both fragmentation and non-EU dependencies,"[11] and many political leaders have shifted their rhetoric away

---

[6] Lucia Retter, Julia Muravska, Ben Williams, and James Black, *Persistent Challenges in UK Defence Equipment Acquisition*, RAND Corporation, RR-A1174-1, 2021.

[7] Elena Lazarou with Alexandra M. Friede, "Permanent Structured Cooperation (PESCO): Beyond Establishment," European Parliamentary Research Service, March 2018.

[8] Max Bergmann, Mathieu Droin, Sissy Martinez, and Otto Svendsen, "The European Union Charts Its Own Path for European Rearmament," Center for Strategic and International Studies, March 8, 2024.

[9] European Commission, 2024a.

[10] European Commission, 2024a.

[11] Jonata Anicetti, "EU Arms Collaboration, Procurement, and Offsets: The Impact of the War in Ukraine," *Policy Studies*, Vol. 45, No. 3–4, May–July 2024, p. 443.

from the previously used language of "European strategic autonomy."[12] Again, adding data from such non-EU countries as Norway and the UK would not significantly alter the overall picture. Indeed, the exclusion of the UK—one of Europe's biggest military powers and defense industrial players—from EU initiatives post-Brexit only further frustrates efforts to achieve a pan-European approach to defense cooperation.

The result is a fragmented EDEM and an EDTIB that—while sizable and internationally competitive in some areas—ultimately fails to punch its weight. The disjointed approach of governments to procurement (demand) compounds the industrial capacity and capability (supply) issues that have arisen over three decades of low investment in defense since the end of the Cold War. Ultimately, the "wicked problem" posed by this complex mix of demand and supply issues causes reduced economies of scale and market forces and, thus, competitiveness at the collective European-NATO level—even if certain national industries continue to benefit from this more fragmented and protectionist approach.[13]

The Role of PPBE Processes in Enabling Cooperation

Making more-efficient use of resources would be a good start. Cooperation is one potential mechanism, though not the only one, and we do not suggest that cooperation is a panacea to all the issues described. However, understanding the extent to which defense PPBE processes across Europe enable, incentivize, or actively discourage cooperation with allies and partners (including the United States) could be a step toward engendering a more coherent approach to the challenges faced by countries on both sides of the Atlantic.

---

[12] Lucia Retter, Stephanie Pezard, Stephen Flanagan, Gene Germanovich, Sarah Grand Clement, and Pauline Paille, *European Strategic Autonomy in Defence: Transatlantic Visions and Implications for NATO, US and EU Relations*, RAND Corporation, RR-A1319-1, 2021.

[13] Nations certainly benefit from protectionist defense industrial policies in terms of short-term support to local firms and jobs, increased military security of supply, and so on. Firms, in turn, can benefit from a de facto monopoly within the defense markets of many countries. However, nations may find that, in the long term, it becomes increasingly costly and difficult to sustain large domestic industries through national budgets alone because of a lack of competition or economies of scale that could help to lower costs and promote innovation. This issue is reflected in the push of many countries toward defense exports as a way of avoiding a monopoly-monopsony relationship between governments and national champions in industry; exports offer an alternative source of revenues to fund R&D, production lines, and workforces. In practice, however, there are only so many potential export customers (especially if many EU and NATO countries remain focused either on national suppliers or on buying U.S. equipment), and therefore, there exists an element of fratricide between EU and NATO countries competing for the same external markets (e.g., in the Middle East). Moreover, at the collective level, a fragmented approach to defense acquisition or industrial policies across the EU or NATO Allies can lead to costly duplication and overcapacity in some areas, undercapacity in others (because orders lack a critical mass or other financial incentives for industry to scale up), and a failure to exploit the comparative advantages of different allied nations (e.g., by exploiting lower labor costs in some countries or niches of technological expertise).

As is the case for European ministries of defense (MoDs), the U.S. Department of Defense's (DoD's) PPBE System is a key enabler for fulfilling DoD's mission. However, given the dynamic threat environment, increasingly capable adversaries, and rapid technological changes, there has been increasing concern in the United States that DoD's resource planning processes are too slow and inflexible to meet warfighter needs.[14] Given this concern, the U.S. Congress mandated the formation of a legislative commission in Section 1004 of the National Defense Authorization Act (NDAA) for Fiscal Year (FY) 2022 and made it responsible for the following tasks:

- examine the effectiveness of the PPBE process and adjacent DoD practices, particularly with respect to defense modernization
- consider potential alternatives to these processes and practices to maximize DoD's ability to respond in a timely manner to current and future threats
- make legislative and policy recommendations to improve such processes and practices for the purposes of fielding the operational capabilities necessary to outpace near-peer competitors, providing data and analytical insight, and supporting an integrated budget that is aligned with strategic defense objectives.[15]

The resultant Commission on PPBE Reform asked RAND's National Defense Research Institute to conduct a study of different approaches to PPBE processes in comparable organizations in order to identify lessons for DoD. This report, a product of that effort, focuses on Europe by drawing on case studies of selected allied and partner nations' approaches to aligning their strategies, plans, and resources to deliver national and collective defense.[16] It offers an overview and analysis of the shared challenges and opportunities that exist for more-coherent approaches to defense PPBE between European countries and the United States.

---

[14] Megan McKernan, Stephanie Young, Andrew Dowse, James Black, Devon Hill, Benjamin J. Sacks, Austin Wyatt, Nicolas Jouan, Yuliya Shokh, Jade Yeung, Raphael S. Cohen, John P. Godges, Heidi Peters, and Lauren Skrabala, *Planning, Programming, Budgeting, and Execution in Comparative Organizations: Vol. 2, Case Studies of Selected Allied and Partner Nations*, RAND Corporation, RR-A2195-2, 2024, p. iii. See also Section 809 Panel, *Report of the Advisory Panel on Streamlining and Codifying Acquisition Regulations*, Vol. 2 of 3, June 2018, pp. 12–13; Brendan W. McGarry, *DOD Planning, Programming, Budgeting, and Execution: Overview and Selected Issues for Congress*, Congressional Research Service, R47178, July 11, 2022, p. 1; and William Greenwalt and Dan Patt, *Competing in Time: Ensuring Capability Advantage and Mission Success Through Adaptable Resource Allocation*, Hudson Institute, February 2021, pp. 9–10.

[15] McKernan, Young, Dowse, et al., 2024, p. iii, citing Public Law 117-81, National Defense Authorization Act for Fiscal Year 2022, December 27, 2021.

[16] These case studies are presented in McKernan, Young, Dowse, et al., 2024; and Stephanie Young, Megan McKernan, Andrew Dowse, Nicolas Jouan, Theodora Ogden, Austin Wyatt, Mattias Eken, Linda Slapakova, Naoko Aoki, Clara Le Gargasson, Charlotte Kleberg, Maxime Sommerfeld Antoniou, Phoebe Felicia Pham, Jade Yeung, Turner Ruggi, Erik Silfversten, James Black, Raphael S. Cohen, John P. Godges, Heidi Peters, and Lauren Skrabala, *Planning, Programming, Budgeting, and Execution in Comparative Organizations: Vol. 5, Additional Case Studies of Selected Allied and Partner Nations*, RAND Corporation, RR-A2195-5, 2024.

Attaining such coherence is about *how* money is allocated and spent, not just *how much*. Clearly, the United States outspends all its allies on defense in absolute terms (and, in most cases, in terms of its share of GDP; see Figure 1.3). The focus of this report and wider body of related RAND research is, therefore, on the ways in which EU and NATO countries are confronting important changes in the strategic environment—with implications for resource allocation decisions—in the context of varying budget sizes, as well as on whether and how various types of cooperation mechanisms are being incorporated into these countries' PPBE processes to promote a more coherent approach to shared challenges.

Of course, different countries do not need to have identical approaches to PPBE to work well together. Such distinct approaches invariably reflect each country's unique history, constitution, governance, and culture. But if allies and partners opt to collaborate on the delivery of new defense capabilities rather than to pursue sovereign alternatives, they should ideally each have governmental processes that actively support and enable such collaboration rather than hindering it. To this end, in Table 1.1, we outline examples from previous RAND research of those features of various nations' approaches to PPBE that can either enable the

FIGURE 1.3

Defense Spending of Selected Countries

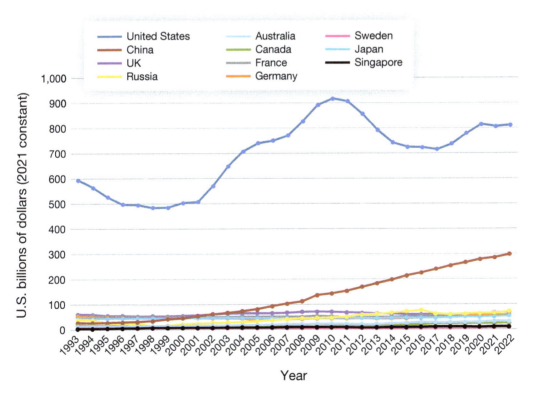

SOURCE: Reproduced from Young et al., 2024, Figure 1.2, using data from Stockholm International Peace Research Institute (SIPRI), "SIPRI Military Expenditure Database," undated. Data shown are as of September 24, 2023.

successful delivery of collaborative programs or inhibit such cooperation in favor of sovereign solutions.

In considering the sorts of potential enablers or barriers of defense cooperation outlined in Table 1.1, we synthesized key trends identified in a subset of allied nations in Europe that were selected for review by the Commission on PPBE Reform. Specifically, we consider the three largest European defense spenders and industrial players in NATO (namely the UK, France, and Germany), as well as Sweden. Besides being the newest NATO member, Sweden deserves consideration as an example of a comparatively small but disproportionately active player industrially, given its long history of armed neutrality (underpinned by an outsized arms industry) to ensure its sovereignty and security of materiel.

TABLE 1.1

PPBE Process Features That Can Work For or Against Defense Cooperation

Activity	Enablers	Barriers
Planning	• Common threat assessment • Similar strategic vision • Similar planning horizon • Alignment of planning process timelines • Participation in cooperation frameworks (e.g., EU, NATO) and shared foresight or requirement-setting activities • Political support for collaboration • Political stability and predictability	• Divergent national policy priorities • Excessive focus on short-term needs over longer-term development of military and industrial capability, capacity, and resilience • Exclusion from cooperation frameworks for political reasons (e.g., Brexit) • Political antagonism (i.e., autarkical or protectionist policies) or unreliability • Sovereignty concerns
Programming	• Mechanisms for the collaborative identification and prioritization of capability requirements (e.g., the International Joint Requirements Oversight Council [I-JROC] in the AUKUS framework) • Willingness to compromise on national requirements • Space within programs for collaboration and coherence across a wider portfolio of collaborative programs[a]	• Lack of timely access to information about other countries' requirements from which to identify opportunities for collaboration • Insistence on bespoke national requirements that preclude, delay, or overcomplicate collaborative programs • Interservice or institutional rivalries • Inability to create space in the wider portfolio of programs for collaborative programs or unwillingness to end or merge national programs (e.g., because of the sunk cost fallacy) • Lack of information, tools, or methods to generate credible business cases, cost estimates, cost-benefit analyses, etc. for collaborative programs

Table 1.1—Continued

Activity	Enablers	Barriers
Budgeting	• Sufficient defense expenditure and acquisition or R&D budgets to fund collaborative programs • Alignment of budgetary cycles • Specific financial mechanisms for incentivizing collaborative programs over national ones (e.g., joint debt, access to EU funds, access to bonus pots of money for industry proposals involving cross-border supply chains) • Access to innovation funds • Access to funding for small- and medium-sized enterprises (SMEs) to take part in collaborative programs • Access to finance and loan facilities (e.g., to support defense exports) • Opportunity for economies of scale	• Insufficient overall levels of defense expenditure, leaving little room for collaborative programs • Prioritization of defense budgets toward personnel, maintenance costs, etc., instead of acquisitions or R&D • Misalignment of budgetary cycles between partners, delaying decisionmaking on joint programs • Lack of long-term budget certainty because of program disruptions caused by domestic politics (e.g., U.S. continuing resolutions, German debt brakes) • Lack of access to central funds or financial mechanisms (e.g., the EU). • Added complexity (and thus costs) of running collaborative programs • Volatility in exchange rates
Execution	• Early agreement on governance and management arrangements • Early agreement on program goals, timelines, etc., and minimization of subsequent changes • Establishment of empowered and effective customer organizations to represent all contributing countries and contract with industry • In absence of such bespoke organizations, use of existing experienced bodies (e.g., OCCAR) • Establishment of equivalent cross-border industrial delivery constructs (e.g., joint ventures among firms) • Acceptance of mutual interdependency and mitigation of sovereignty and security of supply concerns (e.g., ensuring countries can conduct maintenance of critical capabilities even if elements of production take place overseas) • Prioritization of capability, cost, and schedule considerations over a preference for national champions (e.g., opting where possible for the "best athlete" [i.e., the most qualified firm] to deliver a given work package when making workshare allocation decisions) • Minimal duplication of effort (e.g., assembly lines) where possible to control costs	• Preference for national champions in procurement decisions • Invocation of national security exceptions to public procurement law (e.g., blocking use of competition) • Frequent changes to program goals, timelines, etc., and associated disruptions and delays • Insufficiently agile procurement processes and culture of risk aversion • Disagreements over industrial workshare on collaborative programs • Excessive emphasis on *juste retour* rather than "best athlete" approaches to workshare allocation (i.e., allocating industrial workshare based on governments' share of total financial contributions to the program, not firms' competitiveness), undermining the efficiency of collaborative programs and causing delays or quality issues • Disagreements over the handling of intellectual property rights • Tariff and non-tariff barriers to contracting with foreign suppliers • Concerns over intellectual property (IP) and investment security (e.g., attitudes toward Foreign Direct Investment) in partner countries

Table 1.1—Continued

Activity	Enablers	Barriers
Execution, continued	• Early resolution of questions over exports • Efforts to remove red tape affecting cross-border trade (e.g., via trade agreements and government-to-government sales mechanisms) • Alignment of national policies or regulations on IP, security, export, and offsets to minimize friction • Information-sharing mechanisms, including classified information technology (IT) • Trusted relationships and efforts to build a culture of collaboration • Effective risk management • Integration of lessons from previous collaborative programs	• Prohibitive regulations, export controls, etc. (e.g., the United States' International Traffic in Arms Regulations [ITAR]) • Divergent policies on acceptable export customers (e.g., human rights issues) • Impact of national policies and regulations on defense offsets • Insufficient effort to mitigate effects of cognitive bias on program delivery (e.g., optimism bias) • Insufficient risk management or provision • Divergent ways of working and organizational cultures • Language or time-zone barriers • Information-sharing barriers (procedural, cultural, technical) • Lack of institutional memory and failure to learn from previous collaborative programs

SOURCES: Authors' analysis of information from McKernan, Young, Dowse, et al., 2024; Young et al., 2024; Retter, Muravska, et al., 2021.

[a] Collaborative programs can be harder to cut once started because of the added diplomatic fallout of doing so.

Organization of This Report

In Chapter 2, we briefly outline each nation's approach to defense PPBE, drawing on the detailed case studies conducted on behalf of the Commission on PPBE Reform.[17] The discussion then weighs the prospects for achieving greater alignment or coherence among national or collective approaches to PPBE as potential enablers or barriers of wider defense cooperation in Europe. This analysis includes specific consideration of the impact of the collective EU and NATO response to Russia's invasion of Ukraine in February 2022, as well as the spending trends and collaboration mechanisms that have arisen from the West's response to that ongoing conflict. High-profile examples of such mechanisms include efforts to coordinate financial and military aid to Ukraine, the proposals of the EU's new Defence Industrial Strategy for incentivizing greater intra-EU cooperation on defense PPBE, and recent efforts to organize collaborative procurements through NATO agencies.

We then discuss, in Chapter 3, key trends emerging from the four cases considered in the previous chapter with respect to European cooperation on defense PPBE.

Chapter 4 presents our concluding assessment of prospects for refining national, EU, or NATO approaches to PPBE as one way to support and enable more-efficient use of resources to enhance collective defense and deterrence in Europe. Making more-efficient

---

[17] These case studies are presented in McKernan, Young, Dowse, et al., 2024; and Young et al., 2024.

use of resources—whether through collaboration or through other means—would benefit the whole transatlantic community while supporting Ukraine and having positive spillover effects for other regions, such as the Indo-Pacific, where the United States and China engage regularly in strategic competition.

By way of a caveat, it should be noted that the focus of this report is specifically on the role that PPBE processes can play in encouraging or frustrating defense cooperation. This report is not a comprehensive assessment of wider factors contributing to—or constraining—the deepening of defense collaboration across Europe or NATO, nor is it a comparison of the relative merits of collaborative versus sovereign programs as a strategy for various nations. Sovereign programs have their own advantages, of course, in offering governments greater certainty that they can access certain capabilities in times of crisis, and these advantages must be balanced against the benefits and drawbacks of collaboration.

CHAPTER 2

National Approaches to Defense PPBE

In this chapter, we highlight key characteristics and recent developments in the defense PPBE approaches of the UK, France, Germany, and Sweden. These case studies are intended as a cross section of national approaches to defense resourcing *within* each nation to help contextualize the subsequent discussion of barriers and enablers that exist to collaboration *between* nations in Chapter 3.

Table 2.1 offers a summary of national information for the countries presented in this chapter. They represent a selection of founding NATO members and the Alliance's newest entrant (as of March 2024); include a mix of EU and non-EU member states; implement varying defense,

TABLE 2.1

Defense and Political Information for Selected European Countries

	UK	France	Germany	Sweden
Defense budget (USD billions)[a]	68.5	53.6	55.8	7.7
Percentage of GDP spent on defense[a]	2.2	1.9	1.4	1.3
Governmental system	Constitutional monarchy, parliamentary democracy	Constitutional republic; semi-presidential democracy	Federal parliamentary republic	Constitutional monarchy, parliamentary democracy
Defense budget planning cycle	Typically a single-year cycle with contingency for longer-term funding; underpinned by parliamentary scrutiny, five-year strategy documents, and annual publication of ten-year equipment plans	Multiyear planning; president retains wide-ranging defense budgetary planning contingency powers	Annual federal budget with five-year defense plans and special contingency fund	Multiyear long-term strategic financial planning aided by close societal engagement with the defense industry

Table 2.1—Continued

	UK	France	Germany	Sweden
Strategic posture	Historically strong, independent, and expeditionary-focused capability, underpinned by U.S.-UK bilateral relationship; transitioning to middle power with NATO focus but also increased engagement in Indo-Pacific (e.g., via AUKUS) Retains nuclear deterrent.	Historically strong, independent capability, underpinned by investment in sovereign defense industrial base, while seeking European leadership role and influence over EU initiatives Retains nuclear deterrent.	Traditional and historical aversion to significant military spending (pre-2022), but since Russia's full-scale invasion of Ukraine, rapidly shifting strategic culture (e.g., *Zeitenwende* speech); focused on collective NATO-EU defense mechanisms	Focused on societal resilience and model of "total defence," grounded in historical approach to armed neutrality and a strong sovereign defense industrial base Joined Finland in seeking NATO membership (granted in 2024) in response to Russia's invasion of Ukraine in 2022.

SOURCE: Authors' analysis of information from Young et al., 2024.
NOTE: USD = U.S. dollars.
[a] SIPRI, undated. Data are as of March 17, 2023.

acquisition, and industrial policies; and demonstrate a range of differing defense budget management systems and levels of defense expenditure.

Each case study begins by summarizing the country's overall approach to PPBE. Each case study concludes by summarizing the defining features of that country's PPBE processes, along with their associated enablers and barriers to defense collaboration.

United Kingdom

The following case study draws from *Case Studies of Selected Allied and Partner Nations*, Volume 2 of the RAND series of reports on prospective DoD PPBE process reforms.[1] Figure 2.1 offers an overview of the UK. This figure is intentionally more detailed than those that follow for France, Germany, and Sweden because the UK's PPBE processes are no longer covered by those discussed later for EU member states.

Approach to Defense PPBE

The UK spends the second-largest amount on defense among NATO member countries and has closely aligned its military with U.S.-led engagements in recent years, such as those in Iraq, Afghanistan, and Syria and, most recently, the joint operations against Houthi rebels in

---

[1] McKernan, Young, Dowse, et al., 2024.

FIGURE 2.1
Overview of the UK

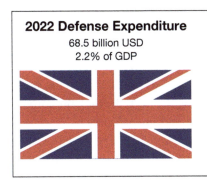

2022 Defense Expenditure	International Alignments	Head of State
68.5 billion USD 2.2% of GDP	• United Nations (UN) Security Council (permanent member) • NATO • Five Eyes • JEF (leader) • Northern Group • EI2 • Group of 7 (G7) • Group of 20 • Exited the EU in 2020	Government is led by the prime minister who controls defense policy and funding with parliamentary scrutiny, with a monarch as ceremonial head of state.

Yemen and the Red Sea. The UK is also the third-biggest donor of military aid to Ukraine, behind the United States and Germany. The UK has played a key role in leading training for Ukrainian forces and has often been the first nation to donate new categories of weapons previously held back in the face of supposed red lines (e.g., main battle tanks, air-launched cruise missiles).

The UK is a constitutional monarchy with a bicameral parliamentary system. The stability of the bicameral system relies on the fact that the chief of the executive branch (the prime minister, formally the First Lord of the Treasury) is a member of parliament from whichever party is able to command the confidence of a majority of the elected lower chamber, the House of Commons. The upper chamber, the House of Lords, is not elected but appointed. Because the UK's government necessarily emerges from the parliament's majority, there is less inherent antagonism between the branches of government than in the United States.

Within this structure, parliament must approve the resources that the UK Ministry of Defence (MOD) requests to perform its mandated military tasks. Funding is typically allocated on a single-year basis, although the MOD currently benefits from a multiyear funding settlement. The UK government is committed to maintaining defense spending above 2 percent of GDP, in line with the NATO target, and aspires to raise its level of defense spending to 2.5 percent once economic conditions permit. Since the Levene reforms of 2011, the UK has structured its MOD in a semi-decentralized manner, which empowers Front Line Commands (the Navy, Army, Air Force, and joint UK Strategic Command) and other top-level budget (TLB) holders (e.g., Defence Nuclear Organisation) with significant delegated authority to prioritize how resources are allocated within their areas of responsibility. These entities are supported by central coordination functions that aim to balance the various parts of the overall portfolio and align strategy with financial cycles—though in practice, there has been debate as to whether the post-Levene shift toward decentralized budgets went too far or not far enough.[2]

Given the MOD's ambitious long-term goals and concurrent requirement to respond to short-term operational pressures, it will need to overcome both internal barriers (e.g., MOD bureaucracy or rivalry between different TLBs) and the destabilizing impact of a confluence of several external trends. In recent years, the UK has experienced an unprecedented period of acute political instability (e.g., it had three prime ministers in 2022) and has faced increased economic and fiscal pressures in the wake of leaving the EU ("Brexit"), the coronavirus disease 2019 (COVID-19) pandemic, and cost-of-living and energy crises.

The war in Ukraine has further tested the flexibility of the UK's budgetary mechanisms in responding to emerging and unplanned requirements. Aid packages to Ukraine have depleted equipment and munition stockpiles. Inflation has been increasing sharply and might force the MOD to cut its budget in real terms despite new funding; it was forced to impose a temporary pause on new capital expenditure from February to March 2024 because of an in-year departmental overrun. Similarly, the defense sector is highly exposed to foreign exchange rate (i.e., trading) volatility, given the extent of its U.S. imports, which are primarily aircraft (e.g., F-35Bs, P-8s, AH-64 Apache helicopters, and CH-47 Chinook helicopters).

The UK's expeditionary focus, international presence through its Overseas Territories, and global commitments require a broad mix of high-end capabilities and the acquisition of equipment for diverse conditions and terrain. This requirement sets the UK apart from other medium powers, such as Germany and Japan, which have narrower defense mission sets and, consequently, a smaller gap between their levels of ambition and resources. Cost growth and escalation challenges have been further intensified by the industrial base and supply chain challenges that the UK has experienced in recent years.

Ongoing Reforms

This mix of long-term and immediate pressures poses significant dilemmas for UK defense planners and those responsible for managing the MOD's finances and executing its spending plans. But these pressures provide added impetus for ongoing efforts to adapt. To this end, the MOD has introduced a variety of measures to encourage innovation, including a new Innovation Fund. And in February 2024, the MOD unveiled a new Integrated Procurement Model, a series of ambitious reforms intended to embrace spiral development and agile acquisition practices as a response to strained resources, a deteriorating threat environment after Russia's invasion of Ukraine, and the accelerating pace of technological development.[3]

---

[2] For additional context on these reforms, see Neil Waghorn, "Lord Levene's Recommendations for Reforming the UK MoD," Defence iQ, June 30, 2011.

[3] UK Ministry of Defence, *Integrated Procurement Model: Driving Pace in the Delivery of Military Capability*, February 2024a.

These reforms include the following:

- a renewed emphasis on managing a defense-wide portfolio and prioritizing requirements centrally to break down organizational stovepipes and encourage integration across the services (in effect, partially reversing the Levene reforms)
- earlier and more-frequent engagement with industry, academia, and science and technology experts in PPBE processes and program-level decisionmaking
- an increased emphasis on exportability as a consideration in program-level decisionmaking to drive economies of scale through multinational cooperation (e.g., joint procurements) and, in turn, boost UK defense industrial resilience
- a deeper partnership with industry and new incentives to drive innovation
- a shift in the MOD's procurement philosophy toward spiral development by default, prioritizing the rapid acquisition of 60- to 80-percent solutions (and then iterating) rather than waiting for the 100-percent solution and an overly exquisite military capability that cannot be fielded at the speed of relevance.

While it is too early to evaluate the impact of this new Integrated Procurement Model, it signals the UK's recognition that PPBE processes urgently need both reform domestically and greater alignment with allies and partners internationally to address the geopolitical, military, technological, industrial, and fiscal pressures currently affecting the MOD. The current UK government (led by the Conservative Party) is not alone in making this assessment. With a general election on July 4, 2024, the opposition Labour Party has similarly set out proposals to reform the MOD's structures and processes. Such plans include reorganizing the MOD as a military strategic headquarters; reforming the Levene model to give the MOD more power over TLBs and drive a more centralized approach to prioritizing capability requirements and investments; empowering UK Strategic Command to assist in this process by driving integration in the face of interservice competition for resources; extending the postings of Service Chiefs from two to four years to create greater leadership continuity; and establishing the new post of National Armaments Director to be responsible for ensuring the delivery of capabilities needed to execute plans and operations.[4]

Thus, there is now cross-party consensus on the need for PPBE reform to address the needs of UK defense in a post–February 2022 world, even if there might still be disagreement on the best way forward for MOD budgets, structures, and processes.

---

[4] John Healey, "A New Era for UK Defence with Labour," video, Policy Exchange, February 28, 2024.

Defining Features of the UK's PPBE Processes, with Collaboration Barriers and Enablers

In summary, the defining features of the UK's approach to defense PPBE are as follows:

- The UK's Defence Operating Model structure and its focus on decentralized decision-making and the empowerment of TLBs in a coordinated but not rigid overarching MOD portfolio seek to balance central oversight and flexibility, empowering Front Line Commands and military end users to determine how best to use their apportioned resources to deliver on their mandated tasks and missions.
- Compared with DoD, the MOD benefits from a lack of legislative intervention in the budgeting process, as well as a three-to-five-year financial settlement that offers some medium-term certainty. The UK then seeks to give some level of certainty to industry partners to guide their investments in R&D, facilities, skills, etc., by publishing an annual ten-year equipment plan, which sets out the MOD's future acquisition plans.
- Mechanisms are in place (e.g., *virement*) to move money between programs or even years if necessary, with either Treasury or parliamentary approval, depending on the circumstances. In practice, however, there are formal and informal (i.e., cultural) barriers to moving money flexibly between different funding buckets, such as departmental expenditure limits for capital expenditure and day-to-day resourcing. These barriers encourage a "use it or lose it" mentality to spending within a given fiscal year, frustrating efforts to run multiannual programs efficiently.
- Despite repeated efforts to cut costs and impose more budgetary discipline, the MOD experiences ongoing challenges in its ability to control cost overruns, and its budget planning remains highly exposed to inflationary pressures and exchange rate volatility—as well as, arguably, a strategic failure to align the government's level of ambition (ends) with its allocated resources (means).[5] The MOD had to pause new capital expenditure for two months in early 2024 as a temporary measure to balance the books for the current fiscal year because of a departmental overrun.[6]
- As in DoD, there are a variety of initiatives underway in the MOD to help drive innovation (including the Innovation Fund and a Defence and Security Accelerator) and a set of mechanisms for agile procurement (e.g., the Urgent Capability Requirement process). Despite these examples of good practice and innovation in program management and delivery, the MOD has assessed that more-ambitious changes are needed to make rapid acquisition the norm rather than the exception and instigated reforms in 2024 to implement a new Integrated Procurement Model.

---

[5] Retter, Muravska, et al., 2021.

[6] Ben Riley-Smith, "MoD Freezes All New Capital Spending as Budgets Spiral Out of Control," *The Telegraph*, February 12, 2024.

The UK's approach to PPBE at the national level in turn influences its approach to multi-national collaboration on defense programs with allies and partners:

- The 2021 Defence and Security Industrial Strategy outlines an "own-collaborate-access" model for capability development and acquisition, along with an emphasis on programs being "international by design."[7] The model sets out what capabilities must absolutely be sovereign (*own*), where the UK can work with allies and partners on joint R&D or procurements to share costs and risk (*collaborate*), and where it can simply buy existing products and services from the open market (*access*).[8]

- The UK is involved in some major multinational development programs, including the A400M airlifter, the Global Combat Air Partnership with Italy and Japan, the European Sky Shield Initiative led by Germany, and the F-35 and Eurofighter programs. It is also a part of OCCAR.

- The UK has embraced an unprecedented depth and breadth of collaboration with the United States and Australia through the AUKUS partnership.[9] This partnership includes new mechanisms for aligning budgets, plans, and capability requirements both bilaterally (e.g., through the Australia-UK ministerial consultations) and trilaterally (e.g., through the I-JROC). Similarly, the UK and Australia benefit from having been designated as "domestic sources" for the U.S. defense industrial base under the FY 2024 NDAA,[10] as well as from efforts to reform U.S. export controls (e.g., ITAR) to enable smoother technology- and information-sharing. Such reforms seek to enable activities under both Pillar I (provision of nuclear-powered submarines to Australia [SSN-AUKUS]) and Pillar II (co-development of disruptive new technologies, such as hypersonic weapons or artificial intelligence [AI]).

- The UK maintains bilateral engagements with individual European nations (including France) under the Lancaster House Treaty, plays a leadership role in the JEF, and contributes to NATO planning processes (such as the NATO Defence Planning Process [NDPP]). The common cause of support to Ukraine has spurred the UK to engage

---

[7] UK Ministry of Defence, *Defence and Security Industrial Strategy: A Strategic Approach to the UK's Defence and Security Industrial Sectors*, March 2021.

[8] Rebecca Lucas, Lucia Retter, and Benedict Wilkinson, *Realising the Promise of the Defence and Security Industrial Strategy in R&D and Exports*, RAND Corporation, PE-A2392-1, November 2022. For more on the implementation of the own-collaborate-access model, see the decision support tool outlined for the MOD in Lucia Retter, James Black, and Theodora Ogden, *Realising the Ambitions of the UK's Defence Space Strategy: Factors Shaping Implementation to 2030*, RAND Corporation, RR-A1186-1, 2022.

[9] Dowse et al., 2024.

[10] Public Law 118-31, National Defense Authorization Act for Fiscal Year 2024, December 22, 2023.

closely with EU member states (such as through a new initiative with Latvia to lead a coalition of countries to fund and produce millions of drones for Ukraine).[11]

- However, since leaving the EU in 2020, the UK is no longer directly engaged in the EU's various mechanisms for aligning defense budgets, plans, or capability development programs.[12] The opposition Labour Party has indicated that, should it win the upcoming election, it would seek to align defense policy and planning more closely with the EU, revitalize the Lancaster House Treaty with France, and pursue a new bilateral treaty with Germany in the first six months in office.[13]

France

The following case study draws from *Additional Case Studies of Selected Allied and Partner Nations*, Volume 5 of the RAND series of reports on prospective DoD PPBE process reforms.[14] Figure 2.2 offers an overview of France.

Approach to Defense PPBE

France is the United States' oldest military ally; it granted financial and military aid to the 13 colonies during the Revolutionary War of 1775–1783. Although France is a founding member of NATO, the nation has pursued a strategy of independence since the end of World War II, particularly prioritizing sovereignty in its military and industrial sectors. France is one of the biggest military spenders in Europe, but it spends far less than the United States and has not reached NATO's target of spending 2 percent of GDP on defense in this century. However, French defense spending has increased steadily over the past ten years, and the French government has committed to reaching the 2-percent goal by 2025.[15]

The French Ministry of Armed Forces (Ministère des Armées, or MinArm) is France's equivalent to DoD and is responsible for protecting French interests both domestically and abroad.[16] Funding is allocated to the MinArm through France's yearly finance law (loi de

---

[11] UK Ministry of Defence, "UK to Supply Thousands of Drones as Co-Leader of Major International Capability Coalition for Ukraine," UK Government, February 15, 2024b.

[12] James Black, Alex Hall, Kate Cox, Marta Kepe, and Erik Silfversten, *Defence and Security After Brexit: Understanding the Possible Implications of the UK's Decision to Leave the EU—Overview Report*, RAND Corporation, RR-1786/1-RC, 2017.

[13] Cristina Gallardo, "UK Labour Would Seek Security and Defense Treaty with Germany," *Politico*, May 16, 2023.

[14] Young et al., 2024.

[15] Vie publique, "Budget de la défense: les étapes pour le porter à 2% du PIB" ["Defense Budget: Steps to Bring It to 2% of GDP"], webpage, December 28, 2022.

[16] MinArm, "Les missions du ministère des Armées" ["The Missions of the Ministry of the Armed Forces"], webpage, undated.

FIGURE 2.2
Overview of France

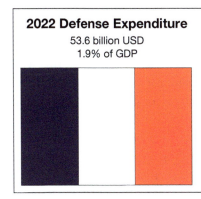

2022 Defense Expenditure	International Alignments	Head of State
53.6 billion USD 1.9% of GDP	• UN Security Council (permanent member) • NATO • EU • G7	Government is led by the president who has direct power over defense policy and funding.

finance), which is informed by the cyclical Military Programming Law (Loi de programmation militaire, or LPM). The LPM determines and defines public spending on defense over a period of four to seven years, which provides some level of medium-term stability. With the LPM serving as the backbone of France's defense budgeting process, the MinArm can then prioritize warfighter and mission needs. The LPM is based on the *Livre blanc sur la défense et la sécurité nationale* (also known as the Strategic Review of Defense and National Security), a strategic document that puts forth the state's defense priorities over several years, as formulated by the French executive branch.

The ability of the French president to settle governmental debates on budget allocation and to enact laws without the need for a parliamentary majority has spared France from the executive-legislative gridlock experienced in the U.S. government. The Strategic Review, the LPM, and the budget breakdown (into missions, programs, actions, subactions, objectives, and performance indicators) provide a coherent framework that encapsulates France's perception of national security threats and its ambitions at the national and international levels. These documents also serve as a military roadmap to fulfilling these ambitions and outline a multiannual budget that decisionmakers can use to enact the changes deemed necessary to support the country's armed forces. The LPM's multiannual reach allows France to plan funding for long-term opportunities and challenges. More generally, France's efforts to maintain independence in defense matters, spearheaded by President Charles de Gaulle in the 1960s, have enabled it to sustain an advanced defense industrial base with a high degree of autonomy from other countries, although it now also seeks to play a leading role within EU-level initiatives to shape the EDTIB and EDEM.

These characteristics come with trade-offs in terms of democratic accountability and adaptability to the evolution of the security environment. In practice, the LPMs tend to evolve in semi-insularity from the rest of the political debate, staying low on the agenda compared with the domestic social positions of candidates during presidential elections every five years.

Defining Features of France's PPBE Processes, with Collaboration Barriers and Enablers

In summary, France's approach to PPBE includes these defining features:

- The French budgetary system ensures certainty in defense spending. France's constitution allows the ruling government to bypass parliamentary opposition on financial matters. The existence of legislative instruments to ensure continuity between budgetary realities and strategic priorities provides further surety for governments in setting defense policy.
- France's technocratic defense culture engenders a close relationship between policy officials and industry and tends to ensure that acquisition decisions are technically grounded. The legislative environment also tends to provide program managers with leeway to ensure budget flexibility through the movement of credits within the budgetary system.
- The flipside of the technocratic coin is that France's PPBE system and its wide-ranging flexibility for officials and the government of the day arguably lack democratic accountability. Legislators are limited in their capacity, in practice, to challenge the government's steering of the national defense strategy and acquisitions. And the close relationship between MinArm officials and the defense industry, as well as an industrial policy focused on support to national champions, creates a reduced role for competition as a means of driving innovation and cost savings compared with the approach taken in the United States and elsewhere.

France's approach to PPBE at the national level in turn influences its approach to multinational collaboration on defense programs with allies and partners:

- France is a leading member of the EU and actively partakes in EU initiatives to promote integration of the EDTIB and EDEM, as well as to coordinate defense planning and capability requirements.
- France has especially close bilateral ties with Germany when it comes to defense, which are reflected in the Aachen Treaty. This treaty commits both nations not only to support one another in the case of an armed attack (augmenting similar clauses in the North Atlantic Treaty and the Treaty of the European Union) but also to develop common strategies for deepening defense cooperation through the EU and to collaborate on defense procurements and exports. This relationship is overseen through the Franco-German Defence and Security Council.[17]

---

[17] Ministère de l'Europe et des Affaires Étrangères, "France Diplomacy: Franco-German Treaty of Aachen," webpage, undated.

- France also collaborates closely with the MOD, especially in the complex weapon sector.[18] Here, the two countries share a close relationship with Anglo-French missile manufacturer MBDA. They collaborate to define and fund shared future requirements for new missiles as part of an Intergovernmental Agreement and the One Complex Weapons program. This agreement explicitly accepts the creation of mutual dependency by having MBDA establish centers of excellence that consolidate certain competencies in France and others in the UK, driving industrial efficiencies and cutting duplication. Other areas of bilateral cooperation include nuclear weapons and the development of unmanned systems and mine countermeasures. The two countries have high-level exchanges through a Defence Ministerial Council and have agreed to coordinate on defense industrial strategy.
- France has driven the creation of the EI2 and plays an active role in NATO, including its planning processes, despite historically seeking to promote a more independent defense policy and support ideas of wider "European strategic autonomy."[19]
- France is actively involved in several flagship multinational programs, such as the A400M airlifter; the Future Combat Air System with Germany, Spain, and Belgium; and the Main Ground Combat System program to develop a new main battle tank with Germany (and possibly others in the future). France is also a member of OCCAR.
- On Ukraine, however, France has been criticized by some NATO Allies on both sides of the Atlantic for giving comparatively less military aid than the UK, Germany, or smaller European nations.[20] France has been reluctant to allow certain EU funding mechanisms to be used to support Ukraine, reflecting occasional wider disagreements between France and the European Commission over the appropriate balance of power between the EU institutions in Brussels and the national governments.[21] However, it has also made recent public statements about the need to increase aid to Kyiv, especially in the face of faltering U.S. support.
- There is an inherent tension in France's approach to defense cooperation. On the one hand, it seeks to deepen the European defense union, bolster the competitiveness of the EDTIB, balance the influence of the United States within the European security architecture, and promote European strategic autonomy through cooperation via the EU and other institutional fora. On the other hand, it maintains a strong emphasis on national sovereignty and independence on defense matters, and its defense industrial policy typically prioritizes French manufacturers over reliance on imports, even if these imports come from European allies. This tension results in a complex and at times even conflic-

---

[18] UK Ministry of Defence, *Equipment Cooperation—United Kingdom and France*, undated.

[19] Retter, Pezard, et al., 2021.

[20] "France and Germany Are at Loggerheads over Military Aid to Ukraine," *The Economist*, February 29, 2024.

[21] "Europe's Damaging Divisions over Military Aid to Ukraine," *Financial Times*, March 5, 2024.

tual relationship between French defense and industrial policies and EU collaboration initiatives (see the discussion of the EU's Defence Industrial Strategy below).[22]

Germany

The following case study draws from *Additional Case Studies of Selected Allied and Partner Nations*, Volume 5 of the RAND series of reports on prospective DoD PPBE process reforms.[23] Figure 2.3 offers an overview of Germany.

Approach to Defense PPBE

Germany's defense budgetary system is shaped by its constitutional arrangements as a federal republic with a bicameral parliamentary system, its role in World War II, and its political and cultural norms around public spending.

Constitutionally, the lower house of the legislature (the Bundestag) is technically more powerful than the upper house (the Bundesrat). The federal government requires the latter's consent only when legislation affects state government revenue or when there is a shift in state and federal responsibilities. German election procedures combine plurality voting with a proportional system. In practice, this process results in a multiparty system in which coalition governments are common, encouraging a political culture focused on consensus-building.

The German military (the Bundeswehr) was established in 1955 as part of the rearmament of West Germany following World War II and the subsequent division of Germany. Under civilian control and overseen by the Federal Ministry of Defense, the Bundeswehr is responsible for defending the country and its interests. Germany's rearmament hinged on its political and military integration into NATO, and the Bundeswehr remains geared toward engagement within a multilateral framework, which nowadays also includes the EU. Therefore, Germany depends on coordination and cooperation with European partners and maintains a close security partnership with the United States.[24]

Germany's budgetary process is highly structured and includes significant checks and balances to ensure that there is sufficient oversight. The theme of accountability to the taxpayer carries throughout its PPBE processes and is firmly grounded in German principles and attitudes toward public spending. Although this careful approach to spending has its drawbacks, Germany has historically held a budget surplus and economically outstrips many of its European counterparts.

---

[22] Jean-Pierre Maulny, *France's Perception of the EU Defence Industrial "Toolbox,"* Armament Industry European Research Group, February 2024.

[23] Young et al., 2024.

[24] Federal Government of Germany, *On German Security Policy and the Future of the Bundeswehr*, June 2016, p. 1.

FIGURE 2.3
Overview of Germany

2022 Defense Expenditure	International Alignments	Head of State
58.8 billion USD 1.4% of GDP	• NATO • EU • Group of 8	Government is led by the Federal Chancellor; the president's role is ceremonial.

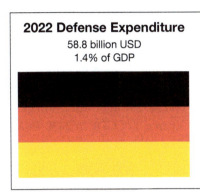

Parliamentary engagement is robust in the German budgetary system, and there is ample opportunity for independent (or semi-independent) institutions to provide input into fiscal policy and budget planning. This approach lends not just increased oversight but also greater validation and diversity of thought and expertise to the budgeting process. Germany issues both an annual budget and a federal financial plan (an internal government planning document covering a five-year period), both of which are renegotiated each year. Beyond the annual federal budget and the five-year financial plan, Germany can tap special funds to finance programs or projects with bespoke objectives and predetermined timelines, which do not need to be renegotiated annually. Historically, special funds were used to support German reconstruction after World War II and infrastructure improvements after reunification in 1990. Introduced in 2009, a *debt brake* limits borrowing to 0.35 percent of GDP per year, except in cases of natural disasters or "extraordinary emergencies beyond the control of the state," as occurred with the COVID-19 pandemic.[25]

Germany has a unique history, which shapes current attitudes toward military spending, as well as public spending more broadly. Nonetheless, Germany is geopolitically located at the heart of Europe, where it plays a core role in ensuring European collective security, particularly against Russia's motives in the region. Although Germany possesses a sizable economy, its defense spending is limited by various political and economic factors, grounded in history, which set it apart from the United States and other allies.

Germany's PPBE mechanisms are similarly unique. The emphasis on accountability and oversight results in unusual arrangements, such as a special fund that allows the Bundeswehr to overcome the debt brake.

---

[25] Steffen Murau and Jan-Erik Thie, *Special Funds and Security Policy: Endowing the German Energy and Climate Fund with Autonomous Borrowing Powers*, Institute for Innovation and Public Purpose, September 2022.

Turning Point

Prior to 2023, Germany's defense plans were articulated every 12 years in white papers that formulated a long-term vision for its security policy and the Bundeswehr. In 2018, the Bundeswehr published a concept that outlined the implementation of the plan described in the most recent (2016) white paper. The objective was to confirm how the Bundeswehr would adapt conceptually from a planning perspective and how it would develop its capabilities and modernize.[26] Defense spending remains far below NATO's target of 2 percent: Germany allocated only 1.2 percent of its GDP to defense even in the years following Russia's annexation of Crimea in 2014. This commitment rose slightly to 1.5 percent of GDP in 2020–2021, as economic output took a hit from COVID-19, but spending had fallen back to 1.4 percent of GDP by the time Russia launched its full-scale invasion of Ukraine in 2022.[27]

The invasion—an early test of Germany's new left-leaning coalition government—prompted an unprecedented and historic shift in German foreign and security policy, known as the *Zeitenwende,* or "turning point."[28] Given Germany's unique history, its political and popular support for defense spending and operations had long been much lower than in many other European nations, not to mention Germany's constitutional restrictions on its use of force and a highly cautious armaments export policy. The events of February 2022 challenged this status quo, prompting both a dramatic shift in energy policy—to wean Germany off its reliance on Russian oil and gas—and an accompanying drive to rebuild German military capabilities and provide support to Ukraine.

In 2023, amid the invasion, the 2016 white paper was superseded by a first-of-its-kind national security strategy. The strategy aimed to "address current and future challenges to our security policy, be they domestic or external, in a comprehensive, integrated and innovative approach which is interministerial and spans all levels."[29] This strategy document was accompanied by Chancellor Olaf Scholz's pledge to meet the NATO target of 2 percent of GDP on defense spending and the establishment of a €100 billion ($107.7 billion) "Zeitenwende special fund" to be written into the country's Basic Law and exempted from the debt brake as a means of urgently addressing shortcomings in German military capabilities and readiness. Immediate spending priorities for the fund include the acquisition of F-35A fighters (to fulfill the dual-capable aircraft role as part of NATO nuclear sharing) and new uncrewed systems. Although the UK was quicker to give military aid to Ukraine in significant volume, Germany has since overtaken the UK to become the second-largest donor behind the United

---

[26] Philipp Hoffmann, "Die Konzeption der Bundeswehr" ["The Concept of the Bundeswehr"], webpage, German Federal Ministry of Defense, August 3, 2018.

[27] SIPRI, undated.

[28] Pia Fuhrhop, "Germany's Zeitenwende and the Future of European Security," Istituto Affari Internazionali, March 6, 2023.

[29] German Federal Ministry of Defense, "National Security Strategy," webpage, undated.

States, providing large numbers of air defense systems, Leopard 2 main battle tanks, and other equipment.[30]

In November 2023, the German federal government faced a budget crisis when the country's constitutional court ruled that another special fund, originally intended for COVID-19 support, could not be repurposed to finance its transition to a green economy—plunging the affordability of the wider budget into question, given the deficit restrictions and debt brake. However, because of its special status, the Zeitenwende fund was exempted from this spending freeze, even while other parts of the Bundeswehr budget were initially affected.[31]

Backed by this fund, German defense spending is expected to jump to 2.1 percent sometime in 2024, exceeding the NATO target for the first time this century and placing Germany on track to be the Alliance's biggest defense spender behind the United States.[32] At the same time, decades of underinvestment mean that much of this new funding is going toward rebuilding lost capability, and questions have been raised about how efficiently such money will be spent, given this spending spree, and whether industry has the capacity to absorb this spike in demand.[33] Looking to the future, there is significant uncertainty about what will happen to federal government budgets once the temporary injection of €100 billion into defense runs out. Germany is expected to have an annual deficit of €56 billion ($61 billion) by 2028 if it keeps its pledge of spending 2 percent of GDP without increasing the Bundeswehr's core budget once the special fund expires.[34]

Defining Features of Germany's PPBE Processes, with Collaboration Barriers and Enablers

In summary, Germany's approach to defense PPBE includes the following defining features:

- German defense policy is undergoing a period of significant change as the country proactively seeks to address long-standing military and industrial capability shortfalls based on a historically rooted national aversion to defense spending. Moves to streamline defense procurement are now a political priority.
- The presence of special funds and budgetary virement or carry-over mechanisms embed flexibility into German PBBE processes. As long as these special funds are accompanied by a clear strategic direction for their use, they can generate momentum for innovation in defense and can be insulated from the debt brake.

---

[30] Kiel Institute for the World Economy, "Ukraine Support Tracker," database, February 16, 2024.

[31] Aurélie Pugnet and Nick Alipour, "Germany Rushes to Assure Allies About Defence Spending," Euroactiv, November 24, 2023.

[32] "German Military Headed for 56-bln-eur Spending Gap in 2028—Spiegel," Reuters, January 31, 2024.

[33] Harper Fine and Peter Carlyon, "Germany's New Plans for Transforming Its Defence and Foreign Policy Are Bold. They Are Also Running Into Familiar Problems," RAND Blog, January 17, 2024.

[34] "German Military Headed for 56-bln-eur Spending Gap in 2028—Spiegel," 2024.

- Germany's focus on scrutiny, accountability, and expert engagement in PPBE processes ensures wide buy-in and democratic grounding while preventing the disproportionate influence of single-service interests. This cautiousness, however, can undermine pace in acquisition.
- There are unresolved questions about whether Germany can sustain its recent spike in defense spending once the €100 billion special fund expires and whether this temporary injection of new funds will be spent efficiently.

Germany's approach to PPBE at the national level in turn influences its approach to multinational collaboration on defense programs with allies and partners:

- Germany is a leading member of the EU and actively partakes in EU initiatives to promote integration of the EDTIB and EDEM, as well as to coordinate defense planning and capability requirements.
- Germany has especially close bilateral ties with France when it comes to defense (see the abovementioned Aachen Treaty in the France case study).
- The German defense industry has significant manufacturing capability and technical expertise, especially in such fields as tanks and armored vehicles, naval shipbuilding, radars, electronics, and missiles and munitions. Limited domestic defense spending before 2023 has encouraged many German firms to look to international and pan-European programs as sources of revenue.
- Besides taking part in a series of PESCO projects, Germany is a leading participant in such flagship collaborative programs as the Future Combat Air System with France, Spain, and Belgium and the Main Ground Combat System with France and others. Germany previously co-developed the Tornado and Eurofighter Typhoon combat aircraft with the UK, Italy, and Spain.
- Germany's restrictive arms exports policy has caused significant friction in past and ongoing collaborative programs. For example, Germany has held up the export of Eurofighter Typhoons to Saudi Arabia—to the frustration of partner governments and industry players involved—and questions have been raised about whether such policies will also affect the exportability of the Future Combat Air System.[35]
- Since Russia's 2022 invasion of Ukraine, the German government has sought to revise these restrictions, making it harder to sell arms to nondemocratic countries and those accused of human rights abuses, while facilitating exports to democracies, such as Ukraine or those participating in pan-European programs.[36]

---

[35] Dominic Perry, "Dassault Chief Concerned by Impact of Germany on FCAS Export Sales," Flight Global, July 20, 2023.

[36] "Germany Plans New Arms Exports Rules, Easier Exports to Ukraine—Der Spiegel," Reuters, June 10, 2022.

Sweden

The following case study draws from *Additional Case Studies of Selected Allied and Partner Nations*, Volume 5 of the RAND series of reports on prospective DoD PPBE process reforms.[37] Figure 2.4 offers an overview of Sweden.

Approach to Defense PPBE

Sweden is a parliamentary constitutional monarchy. King Carl XVI Gustaf's position as head of state is purely ceremonial and symbolic.[38] Executive power lies instead with the national cabinet (Regering), consisting of various ministers and headed by the prime minister. The 349-member parliament (Riksdag) holds legislative power in conjunction with the agencies that perform the day-to-day work of the government.

Whereas Swedish government *ministries* are led by cabinet ministers and set policy, Swedish government *agencies* are freestanding entities that enjoy a high degree of autonomy.[39] This distinction between ministries and more hands-off agencies sets Sweden's governance system apart from others in Europe or that of the United States.

Sweden has two politically appointed defense ministers: the Minister for Defence, who is responsible for Sweden's military defense, and the Minister for Civil Defence, who is responsible for crisis preparedness and civil defense. This division of authority reflects Sweden's commitment to *total defense*, an all-of-society concept of national defense that includes military, civil, economic, and psychological elements.[40] The Swedish Armed

FIGURE 2.4
Overview of Sweden

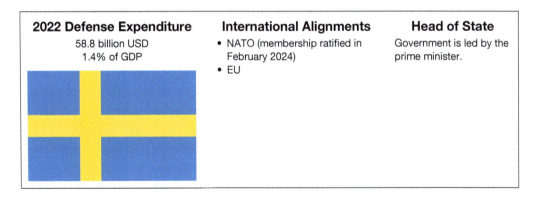

2022 Defense Expenditure	International Alignments	Head of State
58.8 billion USD 1.4% of GDP	• NATO (membership ratified in February 2024) • EU	Government is led by the prime minister.

---

[37] Young et al., 2024.

[38] Government Offices of Sweden, "How Sweden Is Governed," webpage, March 11, 2015.

[39] Swedish Armed Forces, "Organisational Structure and Responsibilities," webpage, last updated April 27, 2023a.

[40] Swedish Ministry of Defence, *Main Elements of the Government Bill Totalförsvaret 2021–2025: Total Defence 2021–2025*, 2020.

Forces are subordinate both to the cabinet via the Minister for Defence and to parliament. The role of the Chief of Defence is to supervise the armed forces and to balance their short-term responses to problems against the country's long-term strategic objectives.[41]

Because of Sweden's focus on total defense, it relies on a system of collective decision-making at the ministerial level—a notable strength of its defense planning and budgeting process. This approach ensures that decisions are made with broad support and buy-in from key stakeholders, which can prevent potential conflicts or disagreements down the line. Additionally, Sweden's tightly regulated and transparent budget cycles promote a clear understanding of the budget process and strategic, informed decisionmaking. The Swedish annual budget is part of a multiannual chain of decisions that is designed to promote financial sustainability.[42] This approach recognizes that budgeting decisions made in one year can have implications for future years and that it is important to take a long-term view when it comes to financial planning on issues of importance to national security and defense.

Long-term financial planning and strategic analysis shape Sweden's forward-looking approach to defense planning that aligns with national objectives. Close partnerships and cooperation between the civilian and military sides, as well as between the state and its defense industry, also help Sweden avoid the U.S. pattern of contentiousness and political gridlock that can lead to uncertainty and inefficiency in the allocation of resources for DoD.

Sweden's approach to materiel supply is closely linked to its total defense concept, whereby endurance is fundamental. The aim is for Sweden to be capable of withstanding a crisis for at least three months.[43] The business sector is instrumental to achieving total defense and security of supply. Despite significant cuts to defense spending (and, consequently, to military capability and readiness) following the end of the Cold War, Sweden maintains a well-developed defense industry with a considerable domestic manufacturing base.

NATO Accession

As in Germany, Russia's invasion of Ukraine in February 2022 provoked a radical shift in Swedish foreign and security policy. Facing a deteriorating threat environment, the Swedish populace expressed strong support for NATO membership in a country that had maintained a policy of armed neutrality for centuries, prompting Sweden to join its Finnish neighbor in applying to join the Alliance in May 2022. Opposition from Türkiye and Hungary to Swedish accession brought a 600-day delay in approving the country's bid; these barriers were overcome in February 2024 with deals for the United States to supply F-16 fighters to Ankara and for Sweden to supply JAS-39 Gripen fighters to Budapest and invest in a joint AI research

---

[41] Swedish Armed Forces, "Chief of Defence," webpage, last updated September 11, 2023b.

[42] Ronnie Downes, Delphine Moretti, and Trevor Shaw, "Budgeting in Sweden," *OECD Journal on Budgeting*, Vol. 2016, No. 2, 2017, pp. 27–28.

[43] Government Offices of Sweden, *Materielförsörjningsstrategi: För vår gemensamma säkerhet* [*Materiel Supply Strategy: For Our Common Security*], May 19, 2022.

center with Hungary. The latter deal cleared the path for Sweden to become NATO's 32nd member, opening up a wide variety of opportunities for deeper collaboration and interoperability with its new NATO Allies.[44]

Defining Features of Sweden's PPBE Processes, with Collaboration Barriers and Enablers

Sweden's approach to defense PPBE includes these defining features:

- Sweden's approach to defense PPBE is grounded in a whole-of-society resilience framework, which binds decisionmaking with close industry partnerships and strong and collective civilian-military and executive oversight.
- Transparency and long-term budgetary planning and collective ministerial-level decisionmaking are routine, which encourages accountability and promotes long-term financial planning, creating a degree of insulation from political changes.
- A lack of key performance indicators in military preparedness and a tendency not to undertake follow-up studies after major procurements hinder the Swedish Armed Forces' overall preparedness.
- A mix of variable and fixed pricing models in budgeting creates inconsistencies.
- Ultimately, Sweden's comparatively agile set of PPBE processes does not fully compensate for its lack of long-term capability planning.

Sweden's approach to defense PPBE at the national level in turn influences its approach to multinational collaboration on defense programs with allies and partners:

- Like France and Germany, Sweden is an active member of the EU and partakes in EU initiatives to promote integration of the EDTIB and EDEM, as well as to coordinate defense planning and capability requirements.
- In 1998, Sweden joined France, Germany, Italy, Spain, and the UK in signing a Letter of Intent supporting the restructuring of European defense industries to promote greater integration and efficiency (although, in practice, success with its implementation has been limited).[45]
- Sweden has especially close ties with Denmark, Finland, Iceland, and Norway, which are reflected in their close cooperation through the NORDEFCO framework. Established in 2009, NORDEFCO merged three other cooperation mechanisms, namely the

---

[44] James Black, Charlotte Kleberg, and Erik Silfversten, *NATO Enlargement Amidst Russia's War in Ukraine: How Finland and Sweden Bolster the Transatlantic Alliance*, RAND Corporation, PE-A3236-1, March 2024.

[45] Nick Witney, "Brexit and Defence: Time to Dust Off the 'Letter of Intent?'" European Council on Foreign Relations, July 14, 2016.

Nordic Coordinated Arrangement for Military Peace Support, the Nordic Armaments Cooperation, and the Nordic Supportive Defence Structures.[46]

- Through NORDEFCO, the member countries share information as part of their national PPBE processes and collaborate on joint working groups and programs. They also engage industry together via the Joint Nordic Defence Industry Cooperation Group.
- Sweden is an active member of the UK-led JEF and the Northern Group, although to date, neither of these groups has focused on joint procurements or other measures to align PPBE processes. For the first time, the JEF activated a JEF Response Option in 2023, consisting of a joint operation by JEF nations to conduct naval and air patrols to secure subsea critical infrastructure, after the severing of a subsea data cable to Sweden and energy connectors between Finland and Estonia.[47]
- Sweden has an outsized defense industry for a country of its comparatively small size, reflecting its long history of armed neutrality backed by domestic arms production to ensure the security of supply as part of total defense. Given the limits of domestic demand, this industry has long been oriented toward exports and has sought to use technology and industrial partnerships to help boost export competitiveness. This orientation is reflected in the close collaboration between Sweden's Saab and Brazil's Embraer on the JAS-39 Gripen fighter (designed in Sweden and now coproduced by Brazil), as well as close industrial ties with the UK (e.g., BAE Systems).[48]
- Joining NATO introduces Sweden to a variety of collaboration mechanisms, including the NDPP, and presents multinational opportunities for collaboration on R&D, innovation, and procurement (discussed further in the next chapter).[49]

---

[46] NORDEFCO, "About NORDEFCO," webpage, undated.

[47] Swedish Ministry of Defence, "Sweden to Take Part in JEF Activity to Protect Critical Infrastructure in Baltic Sea," press release, Government Offices of Sweden, November 28, 2023.

[48] Edward G. Keating, Irina Danescu, Dan Jenkins, James Black, Robert Murphy, Deborah Peetz, and Sarah H. Bana, *The Economic Consequences of Investing in Shipbuilding: Case Studies in the United States and Sweden*, RAND Corporation, RR-1036-AUS, 2015.

[49] Black, Kleberg, and Silfversten, 2024.

CHAPTER 3

European Cooperation on Defense PPBE

As shown in the national case studies, approaches to defense PPBE vary significantly across Europe, which might be unsurprising given that unique constitutional arrangements, fiscal and organizational cultures, and strategic rationales underpin each of them. Key trends emerging from these four cases are as follows:

- Increases in defense expenditure reflect the deterioration in the threat environment, but national MoDs, militaries, and defense industries are all struggling to rebuild capacity at sufficient pace after decades of comparatively low investment.
- Ongoing reforms to national PPBE processes encourage more-agile acquisition of new military capabilities, but there remain lingering questions over whether these reforms will have the desired effect given barriers to change.
- Various new innovation funds and accelerators to pull emerging technologies into defense at a greater pace must compete with enduring barriers, such as a risk-averse bureaucratic culture within MoDs or difficulty accessing financing to enable the commercialization and scaling up of disruptive ideas.
- Like the United States, these European countries strongly prefer to use defense procurements to support domestic industry in areas in which sovereignty and security of supply are deemed paramount.
- These European countries participate in a wide variety of bilateral and multilateral initiatives to develop their own capabilities through collaborative R&D and acquisition programs. There is growing recognition that sovereign programs may no longer be realistic or affordable to fund in as many capability areas as in the past, which increases the incentive to cooperate with like-minded allies to secure benefits for both the military and industry.

These national approaches to defense PPBE in turn shape how European countries approach partnering with others on the resourcing and delivery of defense programs. While the EU has grown significantly in its relevance and military level of ambition as a defense actor since the UK's departure from the EU, defense cooperation in Europe is still addressed primarily at intergovernmental levels—shaped by national initiatives and bilateral or multilateral cooperation between states—rather than supranationally as are other areas of EU policy (e.g., the European Single Market).[1] As a result, the European defense landscape features a patchwork of cooperation, integration, and resource-pooling and resource-sharing initiatives that have been marked by varying degrees of success. Such initiatives include

- EU-led initiatives
- NATO-led initiatives
- bilateral and multilateral initiatives outside the auspices of either the EU or NATO, such as U.S. financing mechanisms open to European allies.

The following sections focus primarily on the first two of these three categories of initiatives because they represent the most significant, mature, and inclusive frameworks for defense cooperation and those with the biggest impact on PPBE processes. We first describe EU-led initiatives, then NATO-led initiatives, before considering the evolution of EU-NATO collaboration and, briefly, U.S. financing mechanisms that support European collective defense. In the final chapter, we assess progress toward deeper defense cooperation against the urgency of action compelled by Russia's full-scale invasion of Ukraine in 2022.

EU-Led Initiatives

Toward a European Defense Union

The EU is a 27-member supranational political and economic bloc, the central policy tenet of which is a single market for goods, services, people, and capital. In joining the EU, member states partially pool their sovereignty and defer policy and regulatory responsibility in a variety of areas in exchange for the benefits of access to a large single market and economies of scale in budgetary matters. The UK left the EU in 2020 following a referendum vote in 2016; as of 2024, it is the only country to do so in the EU's history.

Historically, the roles of the EU and its predecessor institutions in defense have been, by design, somewhat limited. Opt-outs on defense industrial integration were an early feature of the 1957 Treaty of Rome. The initial focus of the European Coal and Steel Community, the EU's predecessor organization, was on post–World War II peacebuilding through the creation of a complex interdependence in key defense materiel sectors—and, therefore, a limited focus on collective defense capability. This history, along with U.S. involvement in NATO, has kept the focus on the NATO Alliance as the guarantor of collective security in Europe.

Nonetheless, the progressive development of the EU's defense and security remit in recent decades has been notable. The establishment of a Common Foreign and Security Policy under the Maastricht Treaty of 1992 and the formation of the European Defence Agency (EDA) under the Lisbon Treaty of 2009 both contributed fundamentally to the EU's rise as an

---

[1] Lucia Retter and Stephanie Pezard, "Rethinking the EU's Role in Collective Defence," *RAND Blog*, May 20, 2022.

"unexpected defense actor."[2] Increasingly galvanized by collective security threats, including most recently Russia's war of aggression in Ukraine, the EU's role in defense is increasing, and the willingness and capacity of the 27 members of the bloc to engage in shared defense initiatives appears to be growing.[3]

The UK's exit from the EU was significant for EU-level defense integration on two fronts: Although the EU lost a significant defense spender and military power, it also lost a member state that had often blocked collective EU defense integration out of fears of duplicating or undermining NATO.[4] Therefore, Brexit, combined with a deteriorating threat environment and concern about possible U.S. disengagement from NATO, has helped motivate a significant transition in the EU's defense role in recent years.

This shift has been reflected in institutional changes in Brussels. In 2021, the European Commission established a Directorate-General for Defence Industry and Space, which significantly elevated the role of the defense industry beyond that found within the existing remit of the small EDA. In 2022, the EU also published its first Strategic Compass for Security and Defence. This vision statement provided overarching policy guidance for the EU's ambition to be a more meaningful defense actor, as well as a common assessment of threats facing Europe (although, because the document was prepared before Russia's invasion of Ukraine in February 2022 but not published until March of that year, elements were rendered immediately obsolete).[5]

EU Process for Capability Planning

The EU has adopted a series of initiatives intended to drive greater alignment of defense PPBE processes across its 27 member states in support of this overall policy ambition toward a closer European Defense Union. Pre-Brexit, these efforts focused primarily on aligning defense capability development across the EU, if with few financial incentives to shape national governments' behaviors toward collaborative approaches. Post-Brexit, these efforts have broadened to include funding, financing, and program delivery mechanisms.

The EU's capability planning process is, by the European Parliamentary Research Service's own admission, "difficult to comprehend because it lacks a name and no formal document fully explains it."[6] This difficulty is compounded by the fact that the EU's process for capability planning involves a "multitude of actors," unfolds in a manner that is "neither

---

[2] Pierre Haroche, "The European Defence Fund: How the European Commission Is Becoming a Defence Actor," Institut de Recherche Stratégique de l'École Militaire, Research Paper No. 56, June 2018.

[3] Retter and Pezard, 2022.

[4] Black et al., 2017.

[5] European External Action Service (EEAS), "A Strategic Compass for Security and Defence," webpage, undated.

[6] Sebastian Clapp, "European Capability Development Planning," European Parliamentary Research Service, March 2024, p. 1.

cyclical nor linear," and represents the suboptimal net result of decades of fragmented and gradualist reform efforts rather than a single comprehensive design.[7]

In simple terms, the process has involved two main elements since its pre-Brexit inception:

1. **Capability Development Mechanism (CDM):** Established in 2003 by mandate of the Treaty of the European Union, the CDM is run by the EU Military Staff (EUMS) within the European External Action Service (EEAS) and is intended to identify military requirements based on the EU's military level of ambition and to highlight shortfalls. These requirements are then proposed to the Political and Security Committee and approved by EU member states' ministers of defense and foreign affairs. In practice, however, the CDM has fallen "into obscurity."[8]

2. **Capability Development Plan (CDP):** Established in 2008, the CDP process is run by the EDA on behalf of its participating member states. The CDP draws on the CDM process and considers both short- and long-term future scenarios and their implications for Europe's collective defense capability requirements across all domains and military tasks. These assessments are then accompanied by an analysis of technological and industrial considerations, resulting in a list of priority areas for action and collaboration. The latest CDP (2023) identified 14 such priorities relating to the five military domains (land, air, maritime, cyber, and space) and an additional eight priorities as strategic enablers.

Collaborative Program Execution

Although the EU cannot *compel* national governments to comply with the CDP, multiple processes are in place to *encourage* collaboration among EU member states in delivering their CDPs. Among the most important of these processes is the Coordinated Annual Review on Defence (CARD). CARD provides a yearly audit of the EU defense landscape and progress toward EU capability goals and toward greater defense coherence and collaboration. Other practical tools include the Collaborative Database, a web-based platform developed by the EDA. This platform provides a repository of national defense plans and requirements and thereby aims to help member states "match-make" with prospective partners for collaborative programs.[9]

In 2017, the EU introduced PESCO as a framework for deeper collaboration among 26 of its 27 member states (all but Malta), going further than what could be agreed through the usual consensus decisionmaking processes. Unlike other mechanisms, PESCO involves

---

[7] Clapp, 2024, p. 1.

[8] Clapp, 2024, p. 1.

[9] EDA, "Collaborative Database," webpage, undated.

legally binding (if arguably unenforceable) commitments from participating European states to reach, among other things, the following goals:[10]

- Regularly increase national defense budgets in real terms.
- In the medium term, increase the share of defense investment expenditure to 20 percent of total defense spending to fill known strategic capability gaps identified through the CDP and CARD processes.
- Increase collaboration on joint capability programs, considering collaborative mechanisms as the priority and using an exclusively national approach only if an examination of collaborative potential mechanisms has already been carried out.
- Use the EDA as the European forum for joint capability development and OCCAR as the preferred collaborative program managing organization.
- Increase defense research and technology expenditure to 2 percent of total defense spending.
- Harmonize requirements and support implementation of the CDP and CARD processes, as well as the European Defence Fund (EDF, which we discuss later).
- Undertake various measures to enhance the availability, deployability, and interoperability of forces in support of EU Common Security and Defence Policy (CSDP) missions and the EU's military level of ambition.
- Ensure that all capability programs contribute to a more competitive EDTIB and seek to align industrial and acquisition policies to reduce duplication.
- Present annual updates on progress against national implementation plans.

In practice, PESCO has been established as a treaty-based framework for joint and collaborative capability development and support to CSDP operations and missions. The EDA, EEAS, and EUMS act together as its secretariat. As of March 2024, some 68 projects were underway via PESCO, each with one or more member states in the lead (as "project coordinator") and other states participating or observing. These projects cover the land, air, maritime, cyber, and space domains and enable defense capabilities, training, and facilities.

Budgetary Mechanisms

There has been significant criticism of the levels of engagement by EU member states with the CDP and CARD processes—or the pace of action to address known capability shortfalls, reduce industrial fragmentation, and implement PESCO projects, most of which have shown few concrete outputs to date. The European Council has similarly found that EU member states "have demonstrated little progress regarding the commitment to increase the number

---

[10] PESCO, "Binding Commitments," webpage, undated.

of collaborative defense capability projects and related investment in defense equipment procurement and defense research and technology."[11]

Therefore, more action is needed from national capitals to translate the potential and ambitions of the CDP, CARD, and PESCO into reality. The EU can bring to bear collective budget capacity and coordination to address some of the shortcomings of the fragmented European market. To incentivize greater collaboration and unblock the greater integration of national PPBE processes, defense plans, and industrial bases, the European Commission has introduced a variety of funding mechanisms in recent years to build capacity incrementally:

- **Precursors to the EDF:** The EU undertook a series of pilot and short-term programs in the mid-2010s to test and refine the approach to establishing long-term EU funding for defense collaboration. These programs included the Pilot Project (PP), a €1.4 million ($1.5 million) grant program run by the EDA in 2016; the Preparatory Action for Defense Research (PADR), a €90 million ($100 million) fund for collaborative defense research that ran from 2017 to 2019; and the European Defence Industrial Development Programme, a more ambitious €500 million ($550 million) program that ran from 2019 to 2020 and that, unlike PP or PADR, required co-funding of collaborative projects from national budgets.[12]

- **EDF:** Launched in 2017 and implemented by the European Commission, the EDF provides a central 2021–2027 budget of €8 billion ($9 billion) to incentivize joint R&D and capability programs through co-funding of cross-border defense initiatives. For this period, the EDF has allocated €2.7 billion ($3 billion) for collaborative defense research and €5.3 billion ($6 billion) for collaborative delivery of capability development programs, such as PESCO. The EDF involves a series of financial incentives for national MoDs to work together, such as a bonus system for collaborative projects that involve SMEs. To be eligible for the EDF, a project must involve a consortium from at least three EU member states (or associated countries, currently meaning only Norway) or entities from at least two such countries in the case of projects concerning disruptive technologies. Part of the EDF, the EU Defence Innovation Scheme, provides additional financial mechanisms to support SMEs, start-ups, and nontraditional contributors to defense programs.[13] A network of national focal points helps potential applicants obtain information and advice on how to engage with the EDF.[14]

---

[11] Aurélie Pugnet, "EU's Flagship Defence Cooperation PESCO Struggles to Show Life," Euractiv, May 19, 2023.

[12] Francesco Giumelli and Marlene Marx, "The European Defence Fund Precursor Programmes and the State of the European Market for Defence," *Defence Studies*, Vol. 23, No. 4, December 2023.

[13] EU, "EU Defence Innovation Scheme (EUDIS)," webpage, undated.

[14] European Commission, Defence Industry and Space, "Network of European Defence Fund National Focal Points (NFP)," webpage, undated.

- **European Defence Industry Reinforcement Through Common Procurement Act (EDIRPA):** Introduced in 2023, EDIRPA is a short-term initiative implemented by the European Commission. The initiative, which provides a €300 million ($330 million) instrument for incentivizing joint procurement of military equipment to address urgent needs, serves as a stopgap measure until a broader and more long-term European Defence Industry Programme (EDIP) can be implemented.

- **Act in Support of Ammunition Production (ASAP):** Introduced in 2023, the EU's ASAP seeks to deploy up to €1 billion ($1.1 billion) to help address shortfalls in munition stockpiles and industrial capacity in the context of Europe's continuing support to Ukraine in its war against Russian aggression. ASAP emerged from an earlier collective horizon-scanning process undertaken through EDIRPA. This earlier process identified that munition stockpiles across the EU were critically depleted and presented a collective security risk. ASAP seeks to address this issue through a three-track approach: (1) the transfer of existing stocks to Ukraine as a priority, to be followed by (2) a shared capacity-building exercise to replenish stocks, and then (3) collective efforts to ensure ongoing industrial capacity. The ASAP initiative is underpinned by supply chain mapping to understand collective challenges to industrial ramp-up, while facilitating access to external funds. ASAP is the latest high-profile example of the EU's ability to leverage collective financing and integrate action across its 27 member states to address shared challenges. However, the full extent of ASAP's impact on ramping up production capacity remains to be seen.

In addition, the EU has repurposed existing mechanisms to channel funds to Ukraine or help partially reimburse member states for their donations of weapons and materiel to Kyiv. Most notably, the European Peace Facility (EPF)—a "formerly obscure" off-budget financing instrument established in 2021 to support such partners as the African Union in conflict prevention—has emerged as an important vehicle for joint EU support to Ukraine.[15] There has, however, been significant wrangling among EU member states as to what should be eligible for funding, whether a "buy European" policy should be applied on any purchases, and how national contributions to the fund should be calculated. In March 2024, EU member states agreed to create the Ukraine Assistance Fund within the EPF, allocating €5 billion ($5.5 billion) in new funding for Ukraine—beyond the already €6.1 billion ($6.6 billion) that the EPF has paid out for Ukraine aid since February 2022.[16]

---

[15] Jacopo Barigazzi, "EU Seals Deal to Send Ukraine 1M Ammo Rounds," *Politico*, March 20, 2023.

[16] Jacopo Barigazzi, "EU Cash for Ukraine: The Bloc Agrees on a €5B Weapons Fund," *Politico*, March 13, 2024.

European Defence Industrial Strategy

In March 2024, the European Commission unveiled the first-ever European Defence Industrial Strategy (EDIS), aiming to bolster the capacity and competitiveness of the EDTIB and respond to the acute challenges posed by the urgent need to boost defense production after decades of underinvestment, given the ongoing war in Ukraine.[17]

The EDIS outlines the following actions:[18]

- **Support a more efficient expression of collective defense demand.** This support will be based on existing instruments and initiatives, such as the CDP, CARD, and PESCO, and will be boosted by incentivizing EU member states' cooperation in the procurement phase of defense capabilities. A Defence Industrial Readiness Board will be established as a new joint programming and procurement function, along with a new high-level European Defence Industry Group. A new Structure for European Armament Programme (SEAP) will aim to facilitate cooperation by EU member states.
- **Ensure that national and EU budgets support, with the necessary means, the adaptation of the European defense industry** to reflect the new security context.
- **Secure the availability of all defense products through a more responsive and innovative EDTIB.** Measures are proposed to ensure that the EDTIB has at its disposal what it needs even in crisis periods, thereby increasing the EU's security of supply. The European Commission will undertake steps to encourage investment in responsive production capacities, and new financing will be provided to SMEs and small mid-cap firms through a new Fund to Accelerate Defence Supply Chain Transformation (FAST).
- **Support defense exports across and beyond the EDTIB.** This support includes the preparation of plans for a future European military sales mechanism as an analogue to the U.S. FMS program.
- **Mainstream a defense readiness culture across EU policies and all financial instruments.** This activity calls for a review of the European Investment Bank's (EIB's) lending policy later in 2024, along with wider engagement with the financial sector to better support development of the EDTIB (e.g., given the impact of environmental, social, and governance [ESG] standards in making it harder for many defense firms to secure investments or financing).[19] The European Commission also plans to consider whether and how defense readiness and resilience should best be made with explicit objectives in future relevant EU programs.
- **Team up with NATO and other strategic partners;** such collaboration could include an increase in structured dialogues with NATO, the establishment of an EU Defence Inno-

---

[17] European Commission, "First Ever Defence Industrial Strategy and a New Defence Industry Programme to Enhance Europe's Readiness and Security," press release, March 5, 2024b.

[18] EU, "EDIS: European Defence Industrial Strategy," fact sheet, 2024.

[19] European Commission, Defence Industry and Space, "Study Results: Access to Equity Financing for European Defence SMEs," January 11, 2024.

vation Office in Kyiv, the convening of an EU-Ukraine Defence Industry Forum in 2024, and the encouragement of Ukrainian participation in EU defense industry programs.

These actions are accompanied by a set of performance metrics to help assess member states' progress toward industrial competitiveness and readiness, with members invited to do the following:[20]

- Ensure that the value of intra-EU defense trade represents at least 35 percent of the value of the EU defense market by 2030.
- Make steady progress toward procuring at least 50 percent of the value of the defense procurement budget within the EU by 2030 and 60 percent by 2035.
- Procure at least 40 percent of defense equipment collaboratively by 2030.

The unveiling of the EDIS was accompanied by the announcement of the EDIP. This program provides additional funding to assist in scaling up European industrial capacity and ensuring the supply and availability of critical defense products. The EDIP is intended as a stopgap measure until the next Multiannual Financial Framework (MFF) (the EU's overarching budget), because that is not due for another four years. In the interim, the EDIP will provide €1.5 billion ($1.7 billion) in funds for the period of 2025–2027, with the intention of negotiating significantly more funding (perhaps as high as €100 billion [$110 billion]) as part of the next MFF (starting from 2028).[21]

Collectively, the various EU initiatives described above move the EU closer toward a full set of PPBE processes and associated financial instruments across the capability life cycle, as shown in Figure 3.1.

NATO-Led Initiatives

NATO lacks the economic dimension or supranational financial mechanisms of the EU, given its primary focus as a political and military alliance. Nonetheless, NATO still has an important role to play in portions of PPBE processes, most notably in terms of aligning capability planning and requirements-setting across 32 Allies and driving the standardization of equipment.

NATO Process for Capability Planning

The centerpiece of this effort is the NDPP, which provides a framework for harmonizing Allies' defense planning activities to ensure the efficient development of the capabilities and forces

---

[20] EU, 2024.

[21] Camille Grand, "Opening Shots: What to Make of the European Defence Industrial Strategy," European Council on Foreign Relations, March 7, 2024.

FIGURE 3.1
EU Defense Funds and Initiatives

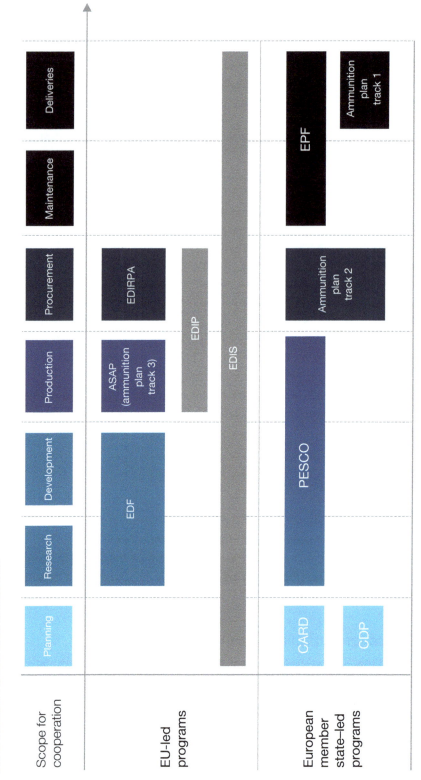

SOURCE: Adapted from Aurélie Pugnet, "Explainer: How to Make Sense of the EU's Defence Funds and Programmes," Euractiv, February 9, 2024.

needed to deliver NATO's collective defense and deterrence. Overseen by the Defence Policy and Planning Committee, the NDPP operates on a four-year cycle and includes sequential activities to provide political guidance, determine requirements, apportion targets to different nations, support implementation, and then review results.[22] This process builds on the overarching vision provided by the NATO 2022 Strategic Concept,[23] aiming to deliver the forces and capabilities needed to operationalize the conceptual framework offered by the Concept for the Deterrence and Defence of the Euro-Atlantic Area and associated regional plans.[24]

Collaborative Program Execution

NATO also plays a growing role in terms of collaborative procurement programs undertaken on behalf of groups of Allies. This role builds on the long-standing work of NATO's Conference of National Armaments Directors (CNAD), which was first established in 1966 and reports directly to the North Atlantic Council. The CNAD helps coordinate and monitor the implementation of capability development decisions undertaken through the NDPP and is supported by the Defence Investment Division.[25] To ensure that industry's perspective is considered, the CNAD is advised by the NATO Industrial Advisory Group, a consultative body of senior industry leaders.[26] The NATO Support and Procurement Agency (NSPA) then helps acquire, operate, and maintain certain capabilities on behalf of groups of Allies or the whole Alliance, historically with a particular focus on such areas as logistics and munitions.[27] The NATO Communications and Information Agency similarly helps acquire and deploy communication and information systems and other related capabilities,[28] and it works closely with industry through the NATO Industry Cyber Partnership.[29]

---

[22] NATO, "NATO Defence Planning Process," webpage, March 31, 2022a.

[23] NATO, "NATO 2022 Strategic Concept," webpage, March 3, 2023b.

[24] NATO, "Deterrence and Defence," webpage, October 10, 2023g.

[25] NATO, "Conference of National Armaments Directors (CNAD)," webpage, January 17, 2023a.

[26] NATO Defence Investment Portal, "NATO Industrial Advisory Group (NIAG)—Introduction," webpage, undated.

[27] NATO, "NATO Support and Procurement Agency (NSPA)," webpage, April 22, 2022b.

[28] NATO Communications and Information Agency, "Who We Are," webpage, undated-b.

[29] NATO Communications and Information Agency, "NATO Industry Cyber Partnership," webpage, undated-a.

At the 2023 Vilnius Summit, the then 31 Allies agreed to a Defence Production Action Plan (DPAP) for NATO, recognizing the acute and urgent production challenges posed by Russia's full-scale invasion of Ukraine. This plan has three themes:[30]

- **Aggregating demand,** including via multinational and multiannual procurements under the aegis of NATO; providing help to Allies in identifying agile procurement and funding mechanisms; and driving an increase in information-sharing on national stockpiles and production requirements. In terms of practical examples, the NSPA recently signed a $5.5 billion contract to procure 1,000 Patriot air defense missiles on behalf of Germany, the Netherlands, Romania, and Spain to help strengthen their defenses and backfill their donated materiel to Ukraine.[31] Through its Land Battle Decisive Munitions framework,[32] the NSPA has similarly put in place framework contracts and around $4 billion of orders to procure a variety of munitions, including 155mm artillery rounds, anti-tank guided missiles, and tank ammunition.[33] Looking beyond the immediate need for munitions, NATO's High Visibility Projects (HVPs) initiative aims to help Allies identify and exploit further opportunities for multinational capability cooperation and development, currently in 21 priority areas.[34]
- **Addressing defense industrial challenges,** including by establishing a new Defence Industrial Production Board, which reports to CNAD and is tasked with convening Allied experts on defense industrial planning and procurement to share best practices. The inaugural meeting of the board in December 2023 spawned three working groups, which aim to address the challenges of understanding the overall industrial capacity (including via the development of relevant metrics), bolstering supply chains, and resolving interoperability and industrial planning–related issues among Allies.
- **Increasing interoperability and standardization,** with the initial focus on the Land Battle Decisive Munitions framework. This element of the DPAP includes activities to increase the visibility of member status and the implementation of NATO standards across the Alliance, improve the materiel standards review process, and further embed NATO standards into the requirements issued to industry by national procurement agencies.

Besides supporting national governments, the Alliance has also invested in acquiring and operating a small number of capabilities centrally. Most notable of these are NATO's fleet

---

[30] NATO, "NATO's Role in Defence Industry Production," webpage, February 12, 2024c.

[31] NATO, "NATO to Buy 1,000 Patriot Missiles to Enhance Allies' Air Defences," January 3, 2024a.

[32] NATO, "NATO Secretary General Welcomes Contracts Worth 2.4 Billion Euros to Strengthen Ammunition Stockpiles," September 28, 2023f.

[33] NATO, "NATO Concludes Contracts for Another $1.2 Billion in Artillery Ammunition," January 23, 2024b.

[34] NATO, "Multinational Capability Cooperation," webpage, March 7, 2024e.

of Airborne Warning and Control System aircraft and the Alliance Ground Surveillance capability, which is based on the RG-4D uncrewed system.[35] Besides its own such programs, NATO also partners with external collaborative frameworks, such as the Strategic Airlift Capability and the Heavy Airlift Wing, based in Hungary, to provide airlift capability to 11 European nations and the United States.[36]

NATO's Drive in Innovation

NATO has also begun in recent years to embrace a more ambitious role in encouraging collaborative R&D and innovation among Allies. This role builds on the work of the NATO Science and Technology Organization and NATO's Science for Peace and Security Program. Prominent new initiatives include the establishment of a €1 billion ($1.1 billion) NATO Innovation Fund—billed as "the world's first multi-sovereign venture capital fund"—and the Defence Innovation Accelerator for the North Atlantic (DIANA).[37] Besides these flagship efforts, the Alliance has also developed a series of technology-related strategies and plans. These include an overarching Strategy for Emerging and Disruptive Technologies, an Artificial Intelligence Strategy, a Strategic Plan for Data, and an Autonomy Implementation Plan. These strategic plans support the work of various stakeholder and expert boards, including the NATO Innovation Board, the Data and Artificial Intelligence Review Board, and the NATO Advisory Group on Emerging and Disruptive Technologies.[38]

EU-NATO Collaboration

The EU's increased role in collective defense should not be seen as duplicating or undermining NATO, if suitably delineated. Indeed, a 2021 RAND study on debates over "European strategic autonomy" concluded that there are myriad benefits to both EU-level and transatlantic security of modernized, more active, and closely integrated EU defense programs. This integration is especially important to bolstering European military and industrial capabilities through the more efficient use of resources via aligned PPBE processes and via pooled and shared initiatives.[39] If done right, NATO and EU defense initiatives should reinforce each other based on each organization's respective areas of strength (e.g., military planning,

---

[35] NATO, "AWACS: NATO's 'Eyes in the Sky,'" webpage, November 14, 2023h; NATO, "Alliance Ground Surveillance (AGS)," webpage, September 4, 2023e.

[36] Allied Air Command, "NATO-Partner Strategic Airlift Capability Provides Airlift to Its 12 Member Nations," North Atlantic Treaty Organization, May 20, 2020.

[37] NATO, "NATO's Innovation Accelerator Becomes Operational and Launches First Challenges," June 19, 2023c.

[38] NATO, "Emerging and Disruptive Technologies," webpage, June 22, 2023d.

[39] Retter, Pezard, et al., 2021.

force generation, operations, and command and control for NATO and defense industrial mobilization for the EU), creating efficiencies and benefits for all. It is telling that the EDIS moves away from the language of "European strategic autonomy" in favor of emphasizing "defense industrial readiness," given the complementarity between many EU and NATO goals in responding to the urgent challenges to military and industrial capacity posed by the war in Ukraine.[40]

In recent years, the EU and NATO have signed several joint declarations seeking to deepen this cooperation between the two entities (reflecting their shared values, heavily overlapping memberships, and largely complementary rather than duplicative areas of strength and focus).[41] The EU and NATO have also emphasized their common stance in relation to their support to Ukraine and bolstering of European defense and industrial capacity—both to address the increased threat from Russia and to ensure burden sharing with the United States.

In practical terms, such political declarations have helped facilitate increased touchpoints between EU and NATO institutions and processes. Efforts have been made to promote coherence between the CARD and NDPP processes, for example, given their overlapping purposes. There has been similar coordination with the EU on NATO's HVPs and an increase in the tempo of consultations between EU and NATO officials more generally.[42]

Additional U.S. Financing Mechanisms for European Allies

In addition to EU and NATO initiatives on defense PPBE, the U.S. government makes several financial mechanisms available to European nations. Two examples are the European Deterrence Initiative, which funds not only the U.S. military's presence and rotations in Europe but also joint exercises and training with European partners, the prepositioning of equipment on the continent, improved infrastructure, and measures to boost European allies' capacities and readiness,[43] and the U.S. Foreign Military Financing program, which provides grants and loans to acquire U.S. military equipment, services, and training either via the FMS program or, in the exceptional cases of Portugal and Greece (as well as eight non-EU countries, such as Israel), via direct commercial contracts.[44]

---

[40] EU, 2024.

[41] European Council, "EU-NATO Cooperation," webpage, undated.

[42] Clapp, 2024.

[43] Office of the Under Secretary of Defense (Comptroller), *European Deterrence Initiative*, U.S. Department of Defense, March 2023.

[44] Defense Security Cooperation Agency, "Foreign Military Financing (FMF)," webpage, undated.

CHAPTER 4

Prospects for Deepening Cooperation via PPBE Process Reform

In this final chapter, we assess progress made toward enabling and incentivizing defense cooperation in Europe through reforms to national or collective PPBE processes. We then consider the implications of persistent barriers, inefficiencies, and resourcing issues for the EU, NATO, and the United States, especially given the urgent need for a coherent collective response to Russia's aggression in Ukraine and other rising threats (e.g., strategic competition with China).

Fresh Impetus: European and U.S. Support to Ukraine

Russia's full-scale invasion of Ukraine in February 2022 has prompted a wide-ranging mobilization of financial, military, and industrial resources across Europe, NATO, and beyond. NATO Allies have collectively contributed around $700 million to the NATO-led Comprehensive Assistance Package Trust Fund for Ukraine. But this sum is dwarfed by bilateral and EU-level donations. According to the Ukraine Support Tracker maintained by the Kiel Institute for World Economy (as of early March 2024), the United States is the biggest single national donor to Ukraine, having provided €68 billion ($75 billion) in total aid, of which €42 billion ($46 billion) is military aid and the rest financial or humanitarian aid. EU institutions (e.g., the European Commission) have provided €77 billion ($84 billion) of aid, although unsurprisingly, given the bloc's remit, its aid has been primarily financial in nature rather than direct military support. Of the case study nations, Germany and the UK have also made significant bilateral donations of military aid to Ukraine, as has Sweden, while France lags far behind (eliciting significant criticism).[1]

Indeed, certain smaller European nations outperform the larger NATO Allies when it comes to giving aid as a percentage of GDP. For example, although the United States is the biggest single donor, it has allocated only 0.32 percent of its GDP to bilateral aid. That ranks it behind 31 other NATO or EU countries by that metric. The leader, Estonia, has offered bilateral aid worth 4.1 percent of its GDP. Collectively, EU member states and their institu-

---

[1] Kiel Institute for the World Economy, 2024.

tions have provided €144 billion ($158 billion) in aid to Ukraine since February 2022, alongside €68 billion ($75 billion) from the United States and €41 billion ($45 billion) from other donors, such as the UK, Japan, Canada, and Australia.[2]

Crucially, as the war continues, Ukraine's supporters have recognized the importance not just of providing more aid but of improving how they provide said aid together. The Ukraine Defense Contact Group's capability coalitions are working to better coordinate funding, production, and delivery of certain categories of equipment. For example, the United States and France are coleading an artillery coalition, Germany is leading an armor capability coalition, and Estonia and Luxembourg are coleading an IT coalition.[3] In June 2024, there have been discussions on whether the United States would retain both its leadership and coordinating roles in the Ukraine Defense Contact Group, with NATO potentially to take on a greater role.[4]

Important Progress but More Needs to Be Done, Urgently

Our analysis has made clear that the war in Ukraine has driven an expansion in multilateral defense collaboration initiatives, often focused on accelerating munition deliveries and production. This urgent need has brought renewed focus on the extent to which national and EU or NATO PPBE-like processes support or undermine efforts to drive deeper cooperation on European defense issues, especially in developing military and industrial capabilities.

Building on the wider RAND research in support of the Commission on PPBE Reform, we have demonstrated in this report that nations on both sides of the Atlantic are grappling with many of the same challenges (though from different national vantage points) and having to tackle such shared problems as how to balance defense spending alongside other priorities (e.g., health or education), how to maximize the efficient use of finite resources, how to drive and absorb innovation from the private sector into the armed forces, and how to procure new capabilities at the speed of relevance, given the rapid pace of evolving threats and technology. The war in Ukraine, and the urgent pressure to ramp up both military and industrial capacity in response, has brought long-standing deficiencies in defense resource management processes into sharper focus. It has also served to stress-test the extent to which such processes enable and incentivize, or hinder, cooperation among allies and partners during a crisis.

Recalling the discussion in Chapter 1 of how PPBE processes can promote or block multinational cooperation (summarized in Table 1.1), we used the ongoing response to the war as a test case of the coherence—or otherwise—between evolving European and U.S. approaches

---

[2] Kiel Institute for the World Economy, 2024.

[3] Joseph Clark, "U.S.-Led Coalition Announces New Initiatives to Bolster Ukraine's Long-Term Armor, Drone Capabilities," *DoD News*, U.S. Department of Defense, January 23, 2024.

[4] Stuart Lau, "NATO to Take Over Part of US-Led Ukraine Aid Channel," *Politico*, June 13, 2024.

to defense PPBE. This assessment shows that the balance has shifted in favor of multinational collaboration (see Table 4.1 at the end of this chapter) in the following ways:

- **The transatlantic community is arguably more closely aligned in its threat assessment and its strategic vision than it has been for decades—at least as they relate to Russia.** There are still significant differences in national assessments of the level of direct threat that Russia poses to NATO Allies, how to handle strategy for the war in Ukraine longer term (including EU and NATO membership), and other national security challenges, such as China, terrorism, migration, instability in the Middle East, and climate change. Nonetheless, the war in Ukraine has refocused European leaders on the importance of a strong national and collective defense and deterrence posture and of having a healthy and resilient industrial base that is able to ramp up production to a war footing when needed. The deteriorating threat environment has also drawn attention on both sides of the Atlantic to the role that PPBE reform can play as a critical—if perhaps unglamorous—enabler of a more robust, agile, adaptable, innovative, and efficient defense enterprise, whether nationally or collectively.

- **National governments, the EU, and NATO have all introduced welcome PPBE reforms and other collaboration initiatives that aim to enhance cooperation to maximize the efficient use of finite resources.** Defense spending has risen across Europe and Canada, increasing at an unprecedented rate of 11 percent in 2023. Two-thirds of NATO Allies are expected to meet or exceed NATO's target of spending 2 percent of GDP on defense in 2024.[5] Of the case study countries, Germany has embarked on an ambitious generational shift in defense spending under the auspices of *Zeitenwende*. Sweden has overhauled its security policy and abandoned neutrality in favor of NATO membership, fully integrating the Nordic-Baltic region (save Kaliningrad) into NATO defense planning processes and initiatives, such as the DPAP, HVPs, and DIANA. France and the UK have both increased their defense spending more cautiously, but the latter is now embarking on major reforms of its approach to defense PPBE through an Integrated Procurement Model that seeks to address some of the inflexibilities exposed by the early lessons of the war in Ukraine. The EU and NATO have similarly been active, introducing such flagship initiatives as EDIRPA, ASAP, and the EDIS, as well as forging the agreed-on DPAP at the Vilnius Summit.

- **Such progress toward greater coherence of national, EU, and NATO approaches to defense PPBE is positive, but countervailing pressures perpetuate a fragmented approach to European defense capability development.** Defense budgets are increasing across Europe—in some countries much quicker than in others—but pressures on TLBs persist amid sluggish economic growth, high inflation, and a cost-of-living crisis. And while the crisis in Ukraine has prompted a surge in rearmament activity across NATO, defense acquisition continues to be highly fragmented in Europe because of countries'

---

[5] NATO, 2024f.

diverse approaches to defense PPBE. These demand-side issues have had a tangible impact on efforts to bolster materiel supply. The EDTIB has been notably slower than the U.S. defense industry to ramp up production, and the EU has fallen short of its production targets for ASAP, which aim to provide 1 million rounds of artillery ammunition to Ukraine.[6] While EU-wide initiatives continue to gain traction, the role of NATO in driving collective European resilience continues to be central, prompting the potential for a further fragmentation of efforts.

- **The focus since February 2022 has been on urgent stopgap measures and immediate procurement needs.** This focus is most evident with EDIRPA or the EDIP, which offer limited short-term funds as a temporary measure until the next MFF. It is also reflected in the strong emphasis on collaborative procurements of lower-tech products (e.g., 155mm artillery rounds) or off-the-shelf munitions (e.g., Patriot air defense missiles), as opposed to the development of newer, high-tech systems to meet future requirements. The war in Ukraine has not had any tangible impact on U.S. or European procurement behaviors when it comes to next-generation combat aircraft, naval vessels, or main battle tanks and armored vehicles, for which the procurement approach remains highly fragmented.

- **There has been less progress toward a more wholesale reassessment of PPBE processes, although the EU's continuing galvanization to become a more robust defense actor holds promise in this area.** The EDIS and its associated EDIP, announced in March 2024, are promising signs of the EU wanting to raise its military level of ambition for closer and more-tangible cooperation on bolstering defense industrial capability and capacity. These developments set some specific metrics for measuring progress in terms of the desired levels of intra-EU trade, collective procurements, and procurements from EU sources by 2030 and 2035. More detail is needed, however, on how specific mechanisms, such as SEAP or FAST, will work in practice, if implementation of the EDIS is to yield more-concrete results than previous EU or PESCO initiatives.[7]

- **There is positive rhetoric on widening the EDTIB's access to private capital and finance, but more needs to be done to make this ambition a reality.** An increasing number of national MoDs have spoken out on the need to address barriers affecting the ability of defense firms, including SMEs, to access venture capital and other forms of funding (especially in the face of ESG rules). The European Commission's proposed changes to EIB's rules on lending is a welcome development toward meeting this need. However, the fact that the commission can only encourage the EIB to lend more, in the EDIS, reflects the limitations on its ability to drive change centrally without buy-in from national governments to changing the EIB's remit. Establishment of the NATO Innovation Fund is a promising move to pool more funds centrally.

---

[6] "EU Will Only Supply Half of Promised Shells to Ukraine by March—Borrell," Reuters, January 31, 2024.

[7] Grand, 2024.

- **While positive messaging about the need to make collaborative programs more efficient is encouraging, few concrete designs have been put forth on how best to achieve such efficiency.** As past RAND studies have shown, multinational defense programs offer significant potential benefits in terms of economies of scale, cost and risk sharing, access to technology, etc. But they also introduce new complexities from a governance and management perspective, including potential misalignment of budgetary cycles or requirements, or disputes over the allocation of industrial workshare. The EDIS's proposal for a new legal framework, SEAP, is welcome (along with associated incentives, such as value-added tax exemptions for any equipment that is jointly owned and procured by multiple nations via a SEAP or the ability of said nations to issue debt through a SEAP). Positive, too, are EU and NATO initiatives to increase information-sharing about existing defense products, national industrial capacities and competencies, and new collaboration opportunities. However, more-detailed information is required on how such ideas as SEAP will work to build confidence that such programs will incorporate best practices and learning from past collaborations. In the meantime, the increased use of experienced agencies, such as OCCAR or NSPA, is encouraging.
- **The war in Ukraine has shown that exports must be viewed more strategically, in the context of building increased industrial readiness across the EU and NATO.** Export sales are important mechanisms for advancing shared goals. Exporters benefit economically, and importers gain access to capabilities that they may not have been able to afford or develop domestically. The EU is taking steps to make government-to-government sales more efficient by scoping an EU foreign sales mechanism as an analogue to the U.S. FMS program, which could have a sizable impact if designed correctly. The EU is similarly seeking to use a system of financial bonuses to incentivize partner countries participating in collaborative programs to agree on export rules and strategy early to avoid situations whereby one nation blocks exports later on (e.g., as happened with Germany blocking Eurofighter sales to Saudi Arabia). The EU is further seeking to smooth intra-EU trade by simplifying such matters as transfer licensing processes. For its part, the UK is seeking to incorporate considerations about exportability earlier into program life cycles as part of its Integrated Procurement Model. The UK also stands to benefit in the future from its own government-to-government system.
- **More clarity is needed on how recent budget increases and PPBE reforms will deliver long-term industrial and supply chain sustainment and resilience.** The war has reignited policy debates over the balance between efficiency and resilience more generally. Dilemmas revolve around how to revise PPBE processes to cultivate a more responsive industrial base that can ramp up production in times of crisis, how to promote security of supply (including friend-shoring with allies), and how to manage the trade-offs between speed, quality, cost, and risk when acquiring innovative technology. The defense industry has repeatedly called for more long-term commitments from national governments to ensure that there is a 5-, 10-, or 15-year demand signal to incentivize private-sector investments in expanding production facilities, workforces, and supply chains for the

future.[8] The new EU grants and funding to support industrial capacity-building have an important role to play in this area. But it is unclear how proposals in the EDIS for add-on orders (based on an "open" regime for defense contracts that enables new customers from within the EU to join on the same terms as old ones) or for European Commission funding of stockpiling will work.[9] There are also more-profound questions to be asked about the legal powers that governments require over their industrial bases in times of crisis, as well as, conversely, how to fund or use slack capacity in times of low demand. Another question is how to resolve political and commercial issues of prioritization when multiple allies are caught in the same queue for new materiel and production rates are limited.

- **Some critical uncertainties need to be resolved, starting with whether the EU can provide the resources needed to deliver on its long-term defense and industrial ambitions.** The provision of only €1.5 billion ($1.7 billion) in new EDIP funds for the period of 2025–2027 represents only 0.2 percent of collective European defense spending. This amount suggests that the EU's flagship program, the EDIS, will be unlikely to have any major short-term impact on shaping the EDTIB or EDEM. While substantial additional funding could be available under the next MFF in 2028, that year will seem too far off to many in the EU, the United States, and Ukraine, given the immediate need to act and fill capability shortfalls during the ongoing war.

- **Another unresolved issue is the relationship between government and industry.** New bodies, such as NATO's Defence Industrial Production Board or the European Defence Industry Group, provide welcome opportunities to feed senior industrialists' perspectives into collective planning and policymaking. More-profound questions persist, however, around government-industry relations. It is notable that the United States has faced fewer issues with its ramp-up of munition production than the EU has, given that the United States unusually has a much less laissez-faire model of industrial policy, compared with many other countries, because U.S. munition factories are government-owned and contractor-operated. Furthermore, the United States has significant emergency powers through its Defense Production Act, which better enables it to compel certain actions from the private sector in a crisis or war—a model replicated in many Nordic-Baltic countries, such as in our case study country of Sweden, but less so in the rest of Europe. While the European Commission has proposed a new toolkit of powers and measures that it can invoke to maintain security of supply in a crisis, this is likely to prove controversial with some EU member states.

- **The relationship between sovereign and collective PPBE mechanisms, including between Brussels and EU member states, remains unresolved.** The constant dilemma

[8] Laura Kayali, Lili Bayer, and Joshua Posaner, "Europe's Military Buildup: More Talk Than Action," *Politico*, June 14, 2023.

[9] Luigi Scazzieri, "The EU's Defence Ambitions Are for the Long Term," Centre for European Reform, March 13, 2024.

of promoting sovereignty versus pooling and sharing resources with allies has been a recurring theme throughout our analysis. The core tension in any collaborative program or effort is to align diverse national approaches to a collective defense PPBE. This is especially relevant to the EDIS, in which the European Commission has proposed a significant expansion of the EU's supranational role (hitherto a national or intergovernmental role) in defense and defense industrial policy. This broader role is likely to receive pushback from some member states, raising questions about the commission's ability to deliver on its full suite of proposals or its ambition for more funds in the next MFF. The Defence Industrial Readiness Board will "only recommend and perhaps cajole," but EU member states will ultimately retain freedom to act as they choose.[10]

- **Another pressing question pertains to the scope for integrating non-EU countries (above all, the UK) into European initiatives.** With the EDIS, the EU has set the precedent of opening many of its defense collaboration opportunities to Ukraine, which is a logical and constructive step, given the vital role that Ukraine's defense now plays in wider European and transatlantic security. Norway can also participate in EDIP as an associated country. However, like the United States, the UK continues to sit outside EU defense initiatives. On an industrial level, this decision excludes some of Europe's leading players in this sector (including the biggest European defense prime contractor, BAE Systems) from meaningful participation in EU programs, given the many third-party restrictions that exist around funding, intellectual property, exports, etc. On a more strategic level, this also means that the UK, one of the biggest defense spenders and military powers in the world and (with France) one of only two nuclear powers in Europe and permanent members of the UN Security Council, sits apart from other European nations.[11] Looking to the longer term, questions will continue to be asked about how to reconfigure the European security architecture post-Brexit and how to best structure EU-UK, EU-U.S., and EU-NATO collaboration on defense.[12]

- **Relatedly, there is some uncertainty around U.S. commitment to Ukraine, the EU, and NATO.** For its part, the United States has struggled in 2023 and early 2024 to maintain its political consensus and legislative support behind aid to Ukraine, raising difficult questions not only for Kyiv but also about transatlantic solidarity and the United States' role in European collective security.[13] Although political systems across Europe are diverse, there is considerably less friction between the executive and legislative branches in defense PPBE in Europe than in the United States. If U.S. support for Ukraine and NATO is to be sustained, redoubled efforts are needed on both sides of the Atlantic to demonstrate the continuing value of such support to U.S. national inter-

---

[10] Scazzieri, 2024.

[11] Black et al., 2017.

[12] Retter, Pezard, et al., 2021.

[13] "America's Political Paralysis Is Complicating Its Support for Ukraine," *The Economist*, December 2, 2023.

ests. Doing so also entails increasing burden sharing by Europeans, as reflected in their recent, sharp uptick in defense spending and provision of aid to Kyiv.

- **Thinking strategically, there is an unresolved question about whether NATO should update its target for defense spending and widen its approach to burden sharing.** For example, NATO could increase the 2 percent of GDP target to 3 percent of GDP or choose to refocus away from this percentage metric onto the actual capabilities delivered, thus emphasizing an output rather than an input. Relatedly, ongoing debates ask whether certain categories of nonmilitary expenditure should be incorporated into metrics of burden sharing. For example, NATO could track how much Allies spend on other areas that contribute to collective security and resilience, such as critical infrastructure, internal security and law enforcement, emergency preparedness, or energy security.[14]
- **The question of political will is a common thread through all these collaboration issues.** Specifically, the question remains whether national governments are willing to resource collective defense in line with the threats faced and whether they are willing to compromise on national sovereignty and short-term industrial returns for longer-term collective benefits. (This question includes the United States' willingness to compromise on its short-term industrial benefits from exporting so much equipment to European governments in favor of building up the EU's and the UK's industrial capacity to the longer-term benefit of NATO and, thus, of the United States as well, especially as it seeks to reprioritize its national defense strategy away from Europe toward competition with China in the Indo-Pacific.) Until such political will to compromise and bear the associated costs happens, progress will be incremental at best and likely insufficient to deal with the scale of the threats faced—not only from Russia and China but also from the risk of U.S. disengagement from European collective security.

Ultimately, then, national, EU, and NATO initiatives to drive collaboration cannot be successful without a change of mindset around defense PPBE in national capitals, while issues of security and defense ultimately remain a sovereign matter. Recent issues, such as Hungary's delaying of EU military and financial aid to Ukraine, have reignited debates about consensus-based decisionmaking versus qualified majority voting on matters of security and foreign policy.[15] It similarly remains to be seen if the European Commission will be able to deliver on the EDIS or its ambition of substantially increased funding for defense through negotiations leading to the next MFF in 2028.

Similarly, a constructive dialogue between the United States and European allies is essential to shape any EU and NATO initiatives in a direction that is beneficial to all parties.[16]

---

[14] Kathleen McInnis, Daniel Fata, Benjamin Jensen, and Jose M. Macias III, "Pulling Their Weight: The Data on NATO Responsibility Sharing," Center for Strategic and International Studies, February 22, 2024.

[15] Centre for Eastern Studies, "The EU Debate on Qualified Majority Voting in the Common Foreign and Security Policy. Reform and Enlargement," October 12, 2023.

[16] Retter, Pezard, et al., 2021.

Regardless of the outcome of the upcoming U.S. presidential election, for which NATO burden sharing and aid to Ukraine have already become divisive topics, European and U.S. officials—including a new NATO Secretary-General—will need to continue to demonstrate their commitment to achieving the shared benefits of transatlantic cooperation on defense and security, including multinational alignment on defense PPBE.[17]

We have shown that the collective response to Russia's invasion of Ukraine continues to drive the reprioritization of defense spending and of new EU, NATO, and bilateral and multilateral collaboration initiatives, but we have not yet found evidence of a more comprehensive effort toward designing a more collective and streamlined approach to defense PPBE. Short of such a holistic effort, piecemeal initiatives are unlikely to achieve true progress toward maximizing the efficient use of resources to bolster collective military and industrial capability and capacity across the EU and NATO. Until then, the EU and NATO will continue to punch below their collective weight.

The war in Ukraine provides both a strategic shock and an opportunity for governments on both sides of the Atlantic to make their respective approaches to defense PPBE more coherent and encourage a more holistic, less fragmented approach to multinational cooperation, ultimately bolstering collective defense, deterrence, and resilience. Timely initiatives currently underway to gather evidence on what does and does not work can inform such efforts. Research by and for the Commission on PPBE Reform,[18] considering the PPBE processes of allied and partner nations,[19] is yielding insights with relevance for both the United States and its European allies and partners.[20] Similarly, ambitious new collaborations outside Europe, such as AUKUS, are driving deeper collaborations across national defense organizations with potentially applicable lessons for the EU.[21]

Informed by such insights, urgent efforts are needed to reform PPBE processes and ensure that national approaches mutually reinforce, rather than undermine, collaborative efforts through the EU and NATO. Only such a coherent approach can enable the transatlantic community to address the scale, pace, and complexity of the strategic challenges arising from national security threats posed by Russia, China, technology disruption, terrorism, and climate change.[22]

---

[17] Franklin D. Kramer and Anca Agachi, "Four NATO Defense Priorities for the Upcoming Washington Summit," *New Atlanticist* blog, Atlantic Council, February 8, 2024.

[18] Commission on Planning, Programming, Budgeting, and Execution Reform, *Defense Resourcing for the Future: Final Report*, U.S. Senate, March 2024.

[19] McKernan, Young, Dowse, et al., 2024.

[20] Megan McKernan, Stephanie Young, Timothy R. Heath, Dara Massicot, Andrew Dowse, Devon Hill, James Black, Ryan Consaul, Michael Simpson, Sarah W. Denton, Anthony Vassalo, Ivana Ke, Mark Stalczynski, Benjamin J. Sacks, Austin Wyatt, Jade Yeung, Nicolas Jouan, Yuliya Shokh, William Shelton, Raphael S. Cohen, John P. Godges, Heidi Peters, and Lauren Skrabala, *Planning, Programming, Budgeting, and Execution in Comparative Organizations: Vol. 4, Executive Summary*, RAND Corporation, RR-A2195-4, 2024.

[21] Dowse et al., 2024.

[22] NATO, 2023b.

TABLE 4.1

PPBE Developments That Favor or Undermine Multinational Cooperation

Activity	Enablers	Barriers
Planning	• Willingness of many European nations to make significant sacrifices in terms of their own military capabilities and stockpiles to aid Ukraine • Sharp increase in the alignment of national governments across the EU and NATO around a common threat assessment (i.e., Russia) • Publication of new strategic visions via the NATO 2022 Strategic Concept and the EU's Strategic Compass (even though the latter was quickly rendered partially obsolete by Russia's invasion of Ukraine) • Publication of industrial strategies (e.g., EDIS) or plans (NATO's DPAP) • Expansion of NATO membership to include Sweden (and presumably Finland) • Expansion of EU defense industrial cooperation to welcome Ukrainian participation • Increased coordination between the EU and NATO on support to Ukraine • Increased political commitment to bolstering both military and defense industrial readiness and resilience in Europe • Emphasis in the EDIS on bolstering the use of CDP and CARD mechanisms to further align defense planning • Interest from many nations in forecasting, scenario planning, net assessment, and wargaming to inform planning for the urgent ramp-up of their defense capabilities	• Continued delays in achieving the necessary political consensus for important and urgent EU or NATO decisions (e.g., Hungary and Türkiye's delaying of Swedish accession to NATO, Hungary's delaying of EU funds for Ukraine, initial French insistence on "buy European" for EPF and opposition to Czechia-led plans to procure 155mm munitions for Ukraine from non-EU sources, and German attempts to have direct aid to Kyiv deducted from national contributions to EPF) • Primary focus on addressing short-term needs (e.g., for more munition deliveries or production), as manifested in 78 percent of new procurements coming from non-European suppliers (especially from the United States), which undermines the longer-term goal of developing European industrial capability, capacity, and resilience • Continued exclusion of third-party countries, such as the UK and, thus, major industrial players (e.g., the biggest European prime contractor, BAE Systems) from EU defense industrial cooperation and funding mechanisms, hampering efforts to achieve a more competitive EDTIB or reduce reliance on the United States • Uncertainty over upcoming U.S. presidential election and the future commitment to NATO of Europe's most important ally, the United States
Programming	• New mechanisms for collaborative identification and prioritization of capability requirements (e.g., via EU's Defence Industrial Readiness Board to promote use of CDP and CARD, or via NATO's new HVPs) • Drive to improve information-sharing (e.g., EU's new catalog of defense products and stockpiles or NATO's push to improve understanding of production capacity across Allies' national industries via its DPAP) • Bilateral and minilateral initiatives to cohere funding and production activities for key capabilities (e.g., first-person view drones)	• Enduring barriers to more-meaningful alignment of national capability development programs, especially for longer-term projects not directly linked to the immediate need to support Ukraine in its predominantly land-centric war (e.g., continued fragmentation of European efforts on next-generation naval or air capabilities) • Interservice disputes over how to use increased budgets or prioritize efforts to bolster readiness, fill long-standing capability gaps, or plug new shortfalls created by donations to Ukraine (e.g., using Germany's special fund)

Table 4.1—Continued

Activity	Enablers	Barriers
Budgeting	• Significant increase in defense expenditure and in procurement budgets across the EU and NATO, which has improved burden sharing with the United States • Some governments' willingness to create special funds (e.g., Germany) or expedite increases to defense budgets outside normal budgetary cycles (e.g., MOD's top-up) • New financial mechanisms for increasing production capacity as an urgent priority (e.g., EU's ASAP) • New financial mechanisms for incentivizing collaborative programs over national ones (e.g., access to financial bonuses via the EDIP or use of the EPF) • Increased access to innovation funds via both the EU and NATO • Increased access to funding for SMEs to take part in collaborative programs and for supply chain transformation (e.g., EU's FAST) • Access to finance and loan facilities (e.g., to support defense exports) and efforts to encourage EIB and the wider financial sector to reform their approaches to lending and investment	• Grossly insufficient funding for EUDIS via EDIP (€1.5 billion) for scale of ambition • Misalignment of EU's budgetary cycles with urgent need to adapt to new strategic threats; next MFF not due to come into force until 2028 • Disparities in national defense spending (e.g., some countries, such as Poland, the Baltic states, and Germany, are moving much more quickly to increase spending than others, such as the UK and France) • Increased fiscal pressures arising from cost-of-living crisis and, especially, efforts by European nations to wean themselves off Russian oil and gas • Increased impact of heightened inflation on defense spending • Prioritization of defense spending toward short-term requirements as opposed to longer-term transformation • Lack of long-term budget certainty, given disruption to programs from domestic politics and the temporary nature of stopgap measures (e.g., unanswered questions over what comes after the German special fund or the initial EDIP funding expire)
Execution	• Increase in off-the-shelf procurements (including via collaborative mechanisms such as NSPA) to simplify delivery • Promotion of SEAP as new model for collaborative programs in the EU, backed by new financial bonuses • Launch of ambitious procurement reforms in some countries (e.g., the UK) • New frameworks for cooperation with industry (e.g., European Defense Industry Group and DPAP) • New frameworks for cooperation with Ukraine (e.g., EU Defense Innovation Office in Kyiv, new EU-Ukraine Defense Industry Forum) • Bonuses via the EDIP and SEAP for early resolution of questions over exports and other matters to speed up programs • Efforts to remove red tape affecting cross-border trade (e.g., via scoping of a new EU military sales mechanism akin to the U.S. FMS program)	• Recent increase in "buying alone and from abroad" to service urgent capability requirements rather than pursuing collaborative programs • Implementation barriers to ambitious national procurement reforms (e.g., in the MOD); feasibility of such reforms • Lack of clarity on how SEAP will work in terms of program management, governance, and delivery arrangements • Lack of concrete new measures to resolve such issues as workshare disputes or delays on new collaborative programs • Limited U.S. ITAR reforms (i.e., the UK, as a partner in AUKUS, benefits but other European partners do not) • Enduring areas of tension in export policies (e.g., Germany's reluctance to export Eurofighter Typhoons to Saudi Arabia) despite some convergence • Enduring information-sharing barriers (procedural, cultural, technical)

NOTE: Our analysis includes PPBE developments only since Russia's invasion of Ukraine in February 2022.

Abbreviations

AI	artificial intelligence
ASAP	Act in Support of Ammunition Production (EU)
AUKUS	Australia–United Kingdom–United States
CARD	Coordinated Annual Review on Defence (EU)
CDP	Capability Development Plan (EU)
CNAD	Conference of National Armaments Directors (NATO)
COVID-19	coronavirus disease 2019
CSDP	Common Security and Defence Policy (EU)
DIANA	Defence Innovation Accelerator for the North Atlantic (NATO)
DoD	U.S. Department of Defense
DPAP	Defence Production Action Plan (NATO)
EDA	European Defence Agency
EDEM	European defence equipment market
EDF	European Defence Fund
EDIP	European Defence Industry Programme
EDIRPA	European Defence Industry Reinforcement Through Common Procurement Act
EDIS	European Defence Industrial Strategy
EDTIB	European defence technological and industrial base
EEAS	European External Action Service (NATO)
EI2	European Intervention Initiative
EIB	European Investment Bank
EPF	European Peace Facility
ESG	environmental, social, and governance
EU	European Union
EUMS	European Union Military Staff
FAST	Fund to Accelerate Defence Supply Chain Transformation (EU)
FMS	Foreign Military Sales (U.S.)
FY	fiscal year
G7	Group of 7
GDP	gross domestic product
HVP	High Visibility Project (NATO)
I-JROC	International Joint Requirements Oversight Council
IP	intellectual property

IT	information technology
ITAR	International Traffic in Arms Regulations (U.S.)
JEF	Joint Expeditionary Force
LPM	Loi de programmation militaire [French Military Programming Law]
MFF	Multiannual Financial Framework (EU)
MinArm	Ministère des Armées [French Ministry of Armed Forces]
MOD	U.K. Ministry of Defence
MoD	Ministry of Defense
NATO	North Atlantic Treaty Organization
NDAA	National Defense Authorization Act (U.S.)
NDPP	NATO Defence Planning Process
NORDEFCO	Nordic Defence Cooperation
NSPA	NATO Support and Procurement Agency
OCCAR	Organisation for Joint Armament Cooperation
PESCO	Permanent Structured Cooperation
PPBE	planning, programming, budgeting, and execution
R&D	research and development
SEAP	Structure for European Armament Programme (EU)
SIPRI	Stockholm International Peace Research Institute
SMEs	small- and medium-sized enterprises
TLB	top-level budget
UK	United Kingdom
UN	United Nations
USD	U.S. dollars

References

Allied Air Command, "NATO-Partner Strategic Airlift Capability Provides Airlift to Its 12 Member Nations," North Atlantic Treaty Organization, May 20, 2020.

"America's Political Paralysis Is Complicating Its Support for Ukraine," *The Economist*, December 2, 2023.

Anicetti, Jonata, "EU Arms Collaboration, Procurement, and Offsets: The Impact of the War in Ukraine," *Policy Studies*, Vol. 45, No. 3–4, May–July 2024.

Barigazzi, Jacopo, "EU Seals Deal to Send Ukraine 1M Ammo Rounds," *Politico*, March 20, 2023.

Barigazzi, Jacopo, "EU Cash for Ukraine: The Bloc Agrees on a €5B Weapons Fund," *Politico*, March 13, 2024.

Bergmann, Max, Mathieu Droin, Sissy Martinez, and Otto Svendsen, "The European Union Charts Its Own Path for European Rearmament," Center for Strategic and International Studies, March 8, 2024.

Black, James, Alex Hall, Kate Cox, Marta Kepe, and Erik Silfversten, *Defence and Security After Brexit: Understanding the Possible Implications of the UK's Decision to Leave the EU—Overview Report*, RAND Corporation, RR-1786/1-RC, 2017. As of March 9, 2024:
https://www.rand.org/pubs/research_reports/RR1786z1.html

Black, James, Charlotte Kleberg, and Erik Silfversten, *NATO Enlargement Amidst Russia's War in Ukraine: How Finland and Sweden Bolster the Transatlantic Alliance*, RAND Corporation, PE-A3236-1, March 2024. As of March 7, 2024:
https://www.rand.org/pubs/perspectives/PEA3236-1.html

Centre for Eastern Studies, "The EU Debate on Qualified Majority Voting in the Common Foreign and Security Policy. Reform and Enlargement," October 12, 2023.

Clapp, Sebastian, "European Capability Development Planning," European Parliamentary Research Service, March 2024.

Clark, Joseph, "U.S.-Led Coalition Announces New Initiatives to Bolster Ukraine's Long-Term Armor, Drone Capabilities," *DoD News*, U.S. Department of Defense, January 23, 2024.

Commission on Planning, Programming, Budgeting, and Execution Reform, *Defense Resourcing for the Future: Final Report*, U.S. Senate, March 2024. As of March 10, 2024:
https://ppbereform.senate.gov/finalreport/

Defense Security Cooperation Agency, "Foreign Military Financing (FMF)," webpage, undated. As of March 10, 2024:
https://www.dsca.mil/foreign-military-financing-fmf

Downes, Ronnie, Delphine Moretti, and Trevor Shaw, "Budgeting in Sweden," *OECD Journal on Budgeting*, Vol. 2016, No. 2, 2017.

Dowse, Andrew, Megan McKernan, James Black, Stephanie Young, Austin Wyatt, John P. Godges, Nicolas Jouan, and Joanne Nicholson, *AUKUS Collaboration Throughout the Capability Life Cycle: Implications for Planning, Programming, Budgeting, and Execution Processes*, RAND Corporation, PE-A2195-1, 2024. As of June 28, 2024:
https://www.rand.org/pubs/perspectives/PEA2195-1.html

EDA—*See* European Defence Agency.

EU—*See* European Union.

"EU Will Only Supply Half of Promised Shells to Ukraine by March—Borrell," Reuters, January 31, 2024.

European Commission, "A New European Defence Industrial Strategy: Achieving EU Readiness Through a Responsive and Resilient European Defence Industry," joint communication to the European Parliament, the Council, the European Economic and Social Committee, and the Committee of the Regions, March 3, 2024a.

European Commission, "First Ever Defence Industrial Strategy and a New Defence Industry Programme to Enhance Europe's Readiness and Security," press release, March 5, 2024b.

European Commission, Defence Industry and Space, "Network of European Defence Fund National Focal Points (NFP)," webpage, undated. As of March 10, 2024: https://defence-industry-space.ec.europa.eu/eu-defence-industry/ network-european-defence-fund-national-focal-points-nfp_en

European Commission, Defence Industry and Space, "Study Results: Access to Equity Financing for European Defence SMEs," January 11, 2024.

European Council, "EU-NATO Cooperation," webpage, undated. As of March 10, 2024: https://www.consilium.europa.eu/en/policies/defence-security/eu-nato-cooperation/

European Defence Agency, "Collaborative Database," webpage, undated. As of March 10, 2024: https://eda.europa.eu/what-we-do/all-activities/activities-search/collaborative-database

European External Action Service, "A Strategic Compass for Security and Defence," webpage, undated. As of March 10, 2024: https://www.eeas.europa.eu/eeas/strategic-compass-security-and-defence-1_en

European Union, "EU Defence Innovation Scheme (EUDIS)," webpage, undated. As of March 10, 2024: https://eudis.europa.eu/index_en

European Union, "EDIS: European Defence Industrial Strategy," fact sheet, 2024.

"Europe's Damaging Divisions over Military Aid to Ukraine," *Financial Times*, March 5, 2024.

Federal Government of Germany, *On German Security Policy and the Future of the Bundeswehr*, June 2016.

Fine, Harper, and Peter Carlyon, "Germany's New Plans for Transforming Its Defence and Foreign Policy Are Bold. They Are Also Running Into Familiar Problems," *RAND Blog*, January 17, 2024. As of March 9, 2024: https://www.rand.org/pubs/commentary/2024/01/germanys-new-plans-for-transforming-its-defence-and.html

"France and Germany Are at Loggerheads over Military Aid to Ukraine," *The Economist*, February 29, 2024.

Fuhrhop, Pia, "Germany's Zeitenwende and the Future of European Security," Istituto Affari Internazionali, March 6, 2023.

Gallardo, Cristina, "UK Labour Would Seek Security and Defense Treaty with Germany," *Politico*, May 16, 2023.

German Federal Ministry of Defense, "National Security Strategy," webpage, undated. As of September 12, 2023: https://www.bmvg.de/en/national-security-policy

"German Military Headed for 56-bln-eur Spending Gap in 2028—Spiegel," Reuters, January 31, 2024.

"Germany Plans New Arms Exports Rules, Easier Exports to Ukraine—Der Spiegel," Reuters, June 10, 2022.

Giumelli, Francesco, and Marlene Marx, "The European Defence Fund Precursor Programmes and the State of the European Market for Defence," *Defence Studies*, Vol. 23, No. 4, December 2023.

Government Offices of Sweden, "How Sweden Is Governed," webpage, March 11, 2015. As of March 1, 2024:
https://www.government.se/how-sweden-is-governed/

Government Offices of Sweden, *Materielförsörjningsstrategi: För vår gemensamma säkerhet* [*Materiel Supply Strategy: For Our Common Security*], May 19, 2022.

Grand, Camille, "Opening Shots: What to Make of the European Defence Industrial Strategy," European Council on Foreign Relations, March 7, 2024.

Greenwalt, William, and Dan Patt, *Competing in Time: Ensuring Capability Advantage and Mission Success Through Adaptable Resource Allocation*, Hudson Institute, February 2021.

Haroche, Pierre, "The European Defence Fund: How the European Commission Is Becoming a Defence Actor," Institut de Recherche Stratégique de l'École Militaire, Research Paper No. 56, June 2018.

Healey, John, "A New Era for UK Defence with Labour," video, Policy Exchange, February 28, 2024. As of March 9, 2024:
https://policyexchange.org.uk/events/a-new-era-for-uk-defence-with-labour/

Hoffmann, Philipp, "Die Konzeption der Bundeswehr" ["The Concept of the Bundeswehr"], webpage, German Federal Ministry of Defense, August 3, 2018. As of September 12, 2023:
https://www.bmvg.de/de/aktuelles/konzeption-der-bundeswehr-26384

Kayali, Laura, Lili Bayer, and Joshua Posaner, "Europe's Military Buildup: More Talk Than Action," *Politico*, June 14, 2023.

Keating, Edward G., Irina Danescu, Dan Jenkins, James Black, Robert Murphy, Deborah Peetz, and Sarah H. Bana, *The Economic Consequences of Investing in Shipbuilding: Case Studies in the United States and Sweden*, RAND Corporation, RR-1036-AUS, 2015. As of February 15, 2024:
https://www.rand.org/pubs/research_reports/RR1036.html

Kiel Institute for the World Economy, "Ukraine Support Tracker," database, February 16, 2024. As of March 9, 2024:
https://www.ifw-kiel.de/topics/war-against-ukraine/ukraine-support-tracker/

Kramer, Franklin D., and Anca Agachi, "Four NATO Defense Priorities for the Upcoming Washington Summit," *New Atlanticist* blog, Atlantic Council, February 8, 2024.

Lau, Stuart, "NATO to Take Over Part of US-Led Ukraine Aid Channel," *Politico*, June 13, 2024.

Lazarou, Elena, with Alexandra M. Friede, "Permanent Structured Cooperation (PESCO): Beyond Establishment," European Parliamentary Research Service, March 2018.

Lucas, Rebecca, Lucia Retter, and Benedict Wilkinson, *Realising the Promise of the Defence and Security Industrial Strategy in R&D and Exports*, RAND Corporation, PE-A2392-1, November 2022. As of February 16, 2024:
https://www.rand.org/pubs/perspectives/PEA2392-1.html

Maulny, Jean-Pierre, *France's Perception of the EU Defence Industrial "Toolbox,"* Armament Industry European Research Group, February 2024.

McGarry, Brendan W., *DOD Planning, Programming, Budgeting, and Execution: Overview and Selected Issues for Congress*, Congressional Research Service, R47178, July 11, 2022.

McInnis, Kathleen, Daniel Fata, Benjamin Jensen, and Jose M. Macias III, "Pulling Their Weight: The Data on NATO Responsibility Sharing," Center for Strategic and International Studies, February 22, 2024.

McKernan, Megan, Stephanie Young, Andrew Dowse, James Black, Devon Hill, Benjamin J. Sacks, Austin Wyatt, Nicolas Jouan, Yuliya Shokh, Jade Yeung, Raphael S. Cohen, John P. Godges, Heidi Peters, and Lauren Skrabala, *Planning, Programming, Budgeting, and Execution in Comparative Organizations:* Vol. 2, *Case Studies of Selected Allied and Partner Nations*, RAND Corporation, RR-A2195-2, 2024. As of January 23, 2024:
https://www.rand.org/pubs/research_reports/RRA2195-2.html

McKernan, Megan, Stephanie Young, Timothy R. Heath, Dara Massicot, Andrew Dowse, Devon Hill, James Black, Ryan Consaul, Michael Simpson, Sarah W. Denton, Anthony Vassalo, Ivana Ke, Mark Stalczynski, Benjamin J. Sacks, Austin Wyatt, Jade Yeung, Nicolas Jouan, Yuliya Shokh, William Shelton, Raphael S. Cohen, John P. Godges, Heidi Peters, and Lauren Skrabala, *Planning, Programming, Budgeting, and Execution in Comparative Organizations:* Vol. 4, *Executive Summary*, RAND Corporation, RR-A2195-4, 2024. As of January 23, 2024:
https://www.rand.org/pubs/research_reports/RRA2195-4.html

MinArm—*See* Ministère des Armées.

Ministère de l'Europe et des Affaires Étrangères, "France Diplomacy: Franco-German Treaty of Aachen," webpage, undated. As of May 17, 2024:
https://www.diplomatie.gouv.fr/en/country-files/germany/france-and-germany/franco-german-treaty-of-aachen/

Ministère des Armées, "Les missions du ministère des Armées" ["The Missions of the Ministry of the Armed Forces"], webpage, undated. As of March 1, 2024:
https://www.defense.gouv.fr/ministere/missions-du-ministere-armees

Murau, Steffen, and Jan-Erik Thie, *Special Funds and Security Policy: Endowing the German Energy and Climate Fund with Autonomous Borrowing Powers*, Institute for Innovation and Public Purpose, September 2022.

NATO—*See* North Atlantic Treaty Organization.

NATO Communications and Information Agency, "NATO Industry Cyber Partnership," webpage, undated-a. As of March 10, 2024:
https://www.ncia.nato.int/business/partnerships/nato-industry-cyber-partnership.html

NATO Communications and Information Agency, "Who We Are," webpage, undated-b. As of March 10, 2024:
https://www.ncia.nato.int/about-us/who-we-are.html

NATO Defence Investment Portal, "NATO Industrial Advisory Group (NIAG)—Introduction," webpage, undated. As of March 10, 2024:
https://diweb.hq.nato.int/niag/Pages_Anonymous/Generic_description.aspx

NORDEFCO—*See* Nordic Defence Cooperation.

Nordic Defence Cooperation, "About NORDEFCO," webpage, undated. As of March 10, 2024:
https://www.nordefco.org/the-basics-about-nordefco

North Atlantic Treaty Organization, "NATO Defence Planning Process," webpage, March 31, 2022a. As of March 10, 2024:
https://www.nato.int/cps/en/natohq/topics_49202.htm

North Atlantic Treaty Organization, "NATO Support and Procurement Agency (NSPA)," webpage, April 22, 2022b. As of March 10, 2024:
https://www.nato.int/cps/en/natohq/topics_88734.htm

North Atlantic Treaty Organization, "Conference of National Armaments Directors (CNAD)," webpage, January 17, 2023a. As of March 10, 2024:
https://www.nato.int/cps/en/natohq/topics_49160.htm

North Atlantic Treaty Organization, "NATO 2022 Strategic Concept," webpage, March 3, 2023b. As of March 10, 2024:
https://www.nato.int/cps/en/natohq/topics_210907.htm

North Atlantic Treaty Organization, "NATO's Innovation Accelerator Becomes Operational and Launches First Challenges," June 19, 2023c.

North Atlantic Treaty Organization, "Emerging and Disruptive Technologies," webpage, June 22, 2023d. As of March 10, 2024:
https://www.nato.int/cps/en/natohq/topics_184303.htm

North Atlantic Treaty Organization, "Alliance Ground Surveillance (AGS)," webpage, September 4, 2023e. As of March 10, 2024:
https://www.nato.int/cps/en/natohq/topics_48892.htm

North Atlantic Treaty Organization, "NATO Secretary General Welcomes Contracts Worth 2.4 Billion Euros to Strengthen Ammunition Stockpiles," September 28, 2023f.

North Atlantic Treaty Organization, "Deterrence and Defence," webpage, October 10, 2023g. As of March 10, 2024:
https://www.nato.int/cps/en/natohq/topics_133127.htm

North Atlantic Treaty Organization, "AWACS: NATO's 'Eyes in the Sky,'" webpage, November 14, 2023h. As of March 10, 2024:
https://www.nato.int/cps/en/natohq/topics_48904.htm

North Atlantic Treaty Organization, "NATO to Buy 1,000 Patriot Missiles to Enhance Allies' Air Defences," January 3, 2024a. As of March 10, 2024:
https://www.nato.int/cps/en/natohq/news_221626.htm

North Atlantic Treaty Organization, "NATO Concludes Contracts for Another $1.2 Billion in Artillery Ammunition," January 23, 2024b.

North Atlantic Treaty Organization, "NATO's Role in Defence Industry Production," webpage, February 12, 2024c. As of March 10, 2024:
https://www.nato.int/cps/en/natohq/topics_222589.htm

North Atlantic Treaty Organization, "Secretary General Welcomes Unprecedented Rise in NATO Defence Spending," February 15, 2024d. As of March 9, 2024:
https://www.nato.int/cps/en/natohq/news_222664.htm

North Atlantic Treaty Organization, "Multinational Capability Cooperation," webpage, March 7, 2024e. As of March 10, 2024:
https://www.nato.int/cps/en/natohq/topics_163289.htm

North Atlantic Treaty Organization, "The Secretary General's Annual Report 2023," webpage, March 14, 2024f. As of March 15, 2024:
https://www.nato.int/cps/en/natohq/opinions_223291.htm

Office of the Under Secretary of Defense (Comptroller), *European Deterrence Initiative*, U.S. Department of Defense, March 2023.

Permanent Structured Cooperation, "Binding Commitments," webpage, undated. As of March 10, 2024:
https://www.pesco.europa.eu/binding-commitments/

Perry, Dominic, "Dassault Chief Concerned by Impact of Germany on FCAS Export Sales," Flight Global, July 20, 2023.

PESCO—*See* Permanent Structured Cooperation.

Public Law 117-81, National Defense Authorization Act for Fiscal Year 2022, December 27, 2021.

Public Law 118-31, National Defense Authorization Act for Fiscal Year 2024, December 22, 2023.

Pugnet, Aurélie, "EU's Flagship Defence Cooperation PESCO Struggles to Show Life," Euractiv, May 19, 2023.

Pugnet, Aurélie, "Explainer: How to Make Sense of the EU's Defence Funds and Programmes," Euractiv, February 9, 2024.

Pugnet, Aurélie, and Nick Alipour, "Germany Rushes to Assure Allies About Defence Spending," Euroactiv, November 24, 2023.

Retter, Lucia, James Black, and Theodora Ogden, *Realising the Ambitions of the UK's Defence Space Strategy: Factors Shaping Implementation to 2030*, RAND Corporation, RR-A1186-1, 2022. As of March 9, 2024:
https://www.rand.org/pubs/research_reports/RRA1186-1.html

Retter, Lucia, Julia Muravska, Ben Williams, and James Black, *Persistent Challenges in UK Defence Equipment Acquisition*, RAND Corporation, RR-A1174-1, 2021. As of March 9, 2024:
https://www.rand.org/pubs/research_reports/RRA1174-1.html

Retter, Lucia, and Stephanie Pezard, "Rethinking the EU's Role in Collective Defence," *RAND Blog*, May 20, 2022. As of March 1, 2024:
https://www.rand.org/pubs/commentary/2022/05/rethinking-the-eus-role-in-european-collective-defence.html

Retter, Lucia, Stephanie Pezard, Stephen Flanagan, Gene Germanovich, Sarah Grand Clement, and Pauline Paille, *European Strategic Autonomy in Defence: Transatlantic Visions and Implications for NATO, US and EU Relations*, RAND Corporation, RR-A1319-1, 2021. As of March 9, 2024:
https://www.rand.org/pubs/research_reports/RRA1319-1.html

Riley-Smith, Ben, "MoD Freezes All New Capital Spending as Budgets Spiral Out of Control," *The Telegraph*, February 12, 2024.

Scazzieri, Luigi, "The EU's Defence Ambitions Are for the Long Term," Centre for European Reform, March 13, 2024.

Section 809 Panel, *Report of the Advisory Panel on Streamlining and Codifying Acquisition Regulations*, Vol. 2 of 3, June 2018.

SIPRI—*See* Stockholm International Peace Research Institute.

Stockholm International Peace Research Institute, "SIPRI Military Expenditure Database," undated. As of September 24, 2023:
https://milex.sipri.org/sipri

Swedish Armed Forces, "Organisational Structure and Responsibilities," webpage, last updated April 27, 2023a. As of September 4, 2023:
https://www.forsvarsmakten.se/en/about/organisation/
organisational-structure-and-responsibilities

Swedish Armed Forces, "Chief of Defence," webpage, last updated September 11, 2023b. As of September 15, 2023:
https://www.forsvarsmakten.se/en/about/organisation/chief-of-defence

Swedish Ministry of Defence, *Main Elements of the Government Bill Totalförsvaret 2021–2025: Total Defence 2021–2025*, 2020.

Swedish Ministry of Defence, "Sweden to Take Part in JEF Activity to Protect Critical Infrastructure in Baltic Sea," press release, Government Offices of Sweden, November 28, 2023.

UK Ministry of Defence, *Equipment Cooperation—United Kingdom and France*, undated.

UK Ministry of Defence, *Defence and Security Industrial Strategy: A Strategic Approach to the UK's Defence and Security Industrial Sectors*, March 2021.

UK Ministry of Defence, *Integrated Procurement Model: Driving Pace in the Delivery of Military Capability*, February 2024a.

UK Ministry of Defence, "UK to Supply Thousands of Drones as Co-Leader of Major International Capability Coalition for Ukraine," UK Government, February 15, 2024b.

Vie publique, "Budget de la défense: les étapes pour le porter à 2% du PIB" ["Defense Budget: Steps to Bring It to 2% of GDP"], webpage, December 28, 2022. As of March 1, 2024:
https://www.vie-publique.fr/eclairage/
284741-budget-de-la-defense-les-etapes-pour-le-porter-2-du-pib

Waghorn, Neil, "Lord Levene's Recommendations for Reforming the UK MoD," Defence iQ, June 30, 2011.

Witney, Nick, "Brexit and Defence: Time to Dust Off the 'Letter of Intent?'" European Council on Foreign Relations, July 14, 2016.

Young, Stephanie, Megan McKernan, Andrew Dowse, Nicolas Jouan, Theodora Ogden, Austin Wyatt, Mattias Eken, Linda Slapakova, Naoko Aoki, Clara Le Gargasson, Charlotte Kleberg, Maxime Sommerfeld Antoniou, Phoebe Felicia Pham, Jade Yeung, Turner Ruggi, Erik Silfversten, James Black, Raphael S. Cohen, John P. Godges, Heidi Peters, and Lauren Skrabala, *Planning, Programming, Budgeting, and Execution in Comparative Organizations: Vol. 5, Additional Case Studies of Selected Allied and Partner Nations*, RAND Corporation, RR-A2195-5, 2024. As of May 8, 2024:
https://www.rand.org/pubs/research_reports/RRA2195-5.html

Milton Keynes UK
Ingram Content Group UK Ltd.
UKHW051413260824
447289UK00016B/2